Hüseyin Koçak

Differential and Difference Equations through Computer Experiments

With Diskettes Containing PHASER: An Animator/Simulator for Dynamical Systems for IBM Personal Computers

Second Edition

With 108 Illustrations

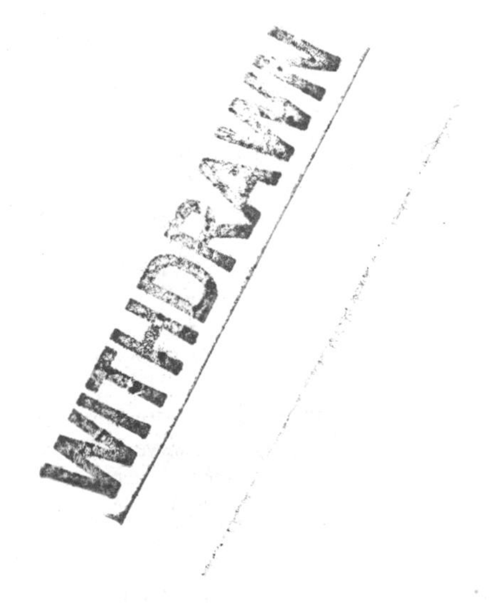

Springer-Verlag
New York Berlin Heidelberg Tokyo

Hüseyin Koçak
Lefschetz Center for Dynamical Systems
Division of Applied Mathematics
Brown University
Providence, RI 02912, U.S.A.

and

Department of Mathematics and
Computer Science
University of Miami
Coral Gables, FL 33124, U.S.A.

AMS Classifications: 34-00, 34A50, 34C05, 34C35

The software enclosed, entitled PHASER, Version 1.1,
utilizes HALO™ graphics sub-routine library.
HALO™ is a trademark of Media Cybernetics, Inc.

Library of Congress Cataloging-in-Publication Data
Koçak, Hüseyin.
Differential and difference equations through computer experiments
: with supplementary diskettes containing PHASER : an
Animator/simulator for dynamical systems for IBM personal computers
/ Hüseyin Koçak. — 2nd ed.
p. cm.
Bibliography: p.
Includes index.
ISBN 0-387-96918-7
1. Differential equations—Computer programs. 2. Difference
equations—Computer programs. I. Title.
QA371.K5854 1988
515.3'5'02885536—dc 19 88-39754

Printed on acid-free paper.

Printed and bound by R.R. Donnelley & Sons, Harrisonburg, Virginia.
Printed in the United States of America.

9 8 7 6 5 4 3 2 1

ISBN 0-387-96918-7 Springer-Verlag New York Berlin Heidelberg
ISBN 3-540-96918-7 Springer-Verlag Berlin Heidelberg New York

To my parents:

Hatice and Irfan,
even though they cannot read these words.

Preface

This is a somewhat unusual book with a dual purpose. First, it is a manual to help readers learn how to use PHASER, the program on the accompanying diskette for IBM personal computers. Second, it is an illustrated guide to the wonderful world of experimental and theoretical dynamics, one which presents dozens of concrete examples ranging from the most rudimentary, appropriate for the beginning student, to the highly complex, suitable for the research mathematician.

Before indicating what PHASER does and how it works, let me describe how it came about. During the past decade the field of differential and difference equations has witnessed a remarkable explosion of knowledge, not only in theory but also in applications to disciplines as diverse as biology and fluid mechanics. Computers have played a crucial role in this process by making possible detailed analyses of specific systems. In this regard, one need only mention the work of Lorenz on strange attractors and the discoveries of Feigenbaum on the bifurcations of interval maps.

It was with the intention of bringing some of this excitement to Brown undergraduates that I began to develop, about three years ago, a new course in the Division of Applied Mathematics bearing the same title as this book. I decided to collaborate with several mathematically-oriented computer science students on the design and the implementation of software for the course in order to take advantage of recent advances in computer-aided instruction by the Department of Computer Science. My main goal was to create a sophisticated interactive simulator for difference and differential equations, one that did not require the user to have any programming knowledge. I also wanted the program to provide the necessary tools for live demonstrations during lectures, and for experimentation and research by students and faculty in Applied Mathematics. After three years, five master's students, and endless revisions, PHASER, and eventually this book, came into being.

Now, let me briefly describe how PHASER works and what it does. It is an extremely versatile and easy-to-use program, incorporating state-of-the-art software technology (menus, windows, etc.) in its user interface. The user first creates, with the help of a menu, a suitable window configuration for displaying a combination of views -- phase portraits, texts of equations, Poincare sections, etc. Next, the user can specify, from another menu, various choices in preparation for numerical computations. He or she can choose, for instance, to study from a library of many dozen equations, and then compute solutions of these equations with different initial conditions or step sizes, while interactively changing parameters in the equations. From yet another menu, these solutions can be manipulated graphically. For example, the user can rotate the images, take sections, etc. During simulations, the solutions can be saved in various ways: as a hardcopy image of the screen, as a printed list, or in a form that can be reloaded into PHASER at a later time for demonstrations or further work.

Since its initial implementation in our computing laboratory two years ago, PHASER has been used in conjunction with our beginning and advanced courses as a means of improving applied mathematics pedagogy through an example-oriented, ''hands-on'' approach. I am happy to be able to say that PHASER has been well received by students and faculty alike at Brown University.

Despite the success of PHASER, I cannot help including a word of caution in this preface. The subject of dynamical systems is vast and colorful, but also inherently difficult. The danger of oversimplification is quite real, and it is easy to get the false impression that computers can provide answers to all questions. The user should always remain aware that numerical simulations have the potential to be misleading. PHASER is not a substitute for the theory of dynamical systems; rather, it is meant to be a complementary tool for performing mathematical experiments and for illustrating the theory with concrete examples. Whatever the purpose for which it is utilized, it is my hope that PHASER will help the subject of ordinary differential and difference equations come alive for its users.

Huseyin Kocak
Brown University
August 1985

Preface to the Second Printing/Edition

Although this constitutes the second printing of *Differential and Difference Equations through Computer Experiments*, the accompanying program is decidedly the second edition, or version 1.1, of PHASER. While the design of the program remains essentially unchanged, it now has the capacity to take advantage of the higher resolution EGA or VGA graphics on IBM Personal Computers. For those who have only CGA graphics, the original version of PHASER is also included. This low resolution version has now been modified to run on EGA/VGA boards as well, primarily for use with inexpensive LCD projectors. For further information regarding installation on a hard disk, printer support, etc., please read the **Read-Me Chapter**, and the `readme` file on the diskette.

I wish to express my gratitude to many users from all continents who have written to me about both PHASER and the text. Your suggestions and critical observations have been most helpful. I only regret not being able to respond to all of you in detail. However, some of the most common requests—a simpler user-interface for entering new equations, a new algorithm (Shampine-Gordon) for stiff systems, and Liapunov exponents, unstable manifolds, basin boundaries, etc.—are slated for inclusion in the projected second edition of the book.

Finally, I would like to thank Springer-Verlag for obtaining a licence to use the HALO graphics library for the development of PHASER.

Hüseyin Koçak
Providence, August 1988

Read-Me Chapter

In this preliminary chapter, we discuss hardware requirements, the contents of chapters, and how to use this book most profitably. As the title suggests, it is required reading for everyone who may wish to use PHASER in instruction or in research.

Whom is PHASER for ?

Modesty aside, *PHASER: An Animator/Simulator for Dynamical Systems* is for all, from freshmen to researchers, who are remotely interested in difference or ordinary differential equations. This book and diskette combination makes a unique complement to the standard textbook approach by allowing the user to analyze specific equations of theoretical and practical importance, and to gain insight into their *dynamics.* In fact, PHASER has been used in the Division of Applied Mathematics, Brown University, both for an elementary course in conjunction with Boyce & DiPrima [1977], and for a more advanced senior/graduate course. The enthusiastic response of our students, above all their return on their own to the computing laboratory to satisfy newly aroused curiosity, has been most pleasing. The sophisticated interactive graphical capabilities of PHASER make it a useful exploratory tool for researchers as well.

Hardware requirements

To run PHASER, you need the following options on your IBM Personal Computer PC, XT, AT, or PS/2:

- DOS Version 2.0 (or higher) operating system,

- 256K-bytes of memory,
- IBM Color Graphics Adapter (CGA), or Enhanced Graphics Adapter (EGA), or Video Graphics Adapter (VGA),
- Color Display. A mono-chrome monitor will work; however, some colors (such as red) may be invisible.
- One version of PHASER supports the Mathematics Coprocessor. We do recommend this hardware option because the gains in computational speeds, especially on a PC or XT, are considerable. The version that requires the Mathematics Coprocessor will not run on a machine that does not have it.

The diskettes

The 5-1/4" diskette in the jacket on the back inside cover is formatted at 1.2M. In the interest of keeping user costs to a minimum, this diskette has been formatted at high density. Those with low density 5-1/4" diskette drives alone will need to transfer the appropriate copy of PHASER to a low density diskette. The two 3-1/2" diskettes, one for EGA/VGA and the other for CGA, are formatted at low density, 720K. The combined contents of these two diskettes are identical to that of the 5-1/4" one.

For safety, make sure the original diskette is write-protected and then make a back-up copy. If you have a hard disk, simply transfer the desired copy of PHASER to the disk, as explained below.

Installation on hard disk

To install PHASER on your hard disk, first make a directory of your choice and change to that directory. Then, insert the appropriate diskette in drive A, and, depending on your hardware configuration, type one of the following install batch commands:

`a:install1` for EGA/VGA graphics WITHOUT Math. Coprocessor
`a:install2` for EGA/VGA graphics WITH Math. Coprocessor
`a:install3` for CGA graphics WITHOUT Math. Coprocessor
`a:install4` for CGA graphics WITH Math. Coprocessor

For example, to install PHASER in a directory named *phaser* on a machine with EGA or VGA graphics and Mathematics Coprocessor, type the following three commands:

`C:> mkdir phaser` < *ENTER* >
`C:> cd phaser` < *ENTER* >

`C:\phaser> a:install2` < *ENTER* >

Now, PHASER is installed in a directory called *phaser*. To run the program, simply type, while in your directory *phaser*:

`C:\phaser> phaser` < *ENTER* >

About ten seconds after you type **phaser**, a logo will appear, followed by the screen image in *Figure 4.2*, and you will be ready to begin simulations.

If you prefer to install PHASER manually, here are the files for EGA or VGA: **phaser.exe** for EGA or VGA graphics WITH Mathematics Coprocessor; **nphaser.exe** for EGA or VGA graphics WITHOUT Mathematics Coprocessor. Copy one of these and the file **demo**. Create a subdirectory named **files** and copy everything in the **files** into it.

All the executables and the supporting files for CGA are in a subdirectory called **cga**. Change to this directory and copy one of the executable files—**phaser.exe** for CGA graphics WITH Mathematics Coprocessor, or **nphaser.exe** for CGA graphics WITHOUT Mathematics Coprocessor—and the remaining files, including everything in the subdirectory **files**.

If you already have PHASER in your computer, copy the desired new executable and the new **files** subdirectory over the existing ones. In this way, you can use your own previously user-entered equations.

Printer support

If you are using the CGA version of PHASER, before starting the program, you should first execute the DOS graphics function by typing

graphics < *ENTER* >

This function allows you to print the contents of the screen on the graphics printer at any time during simulations. This is done by pressing the special <Print Screen>, while holding the <Shift> key.

Unfortunately, the DOS facility <Shift> <Print Screen> does not work with EGA (or VGA) resolution. To remedy this shortcoming of DOS, the current EGA/VGA version of PHASER supports screen dumps for two of the most common printers:

EPSON/IBM dot matrix and *HP LaserJet+* laser printer.

More specifically, when you wish to send the screen to your printer, use the appropriate entry (**E**pson/IBM or **HP** LaserJet+) on the UTILITIES menu. Printing takes several minutes. The aspect ratio of the screen is different from those of the printers; to get the 'correct' aspect ratio on your printer, adjust the *Windosize* from the NUMERICS Menu. In the future, a larger selection of printers will be supported. In the meantime, if you do not have one of these two printers, you may wish to obtain a commercially available memory-resident program for screen dumps.

Notes

- The aspect ratio of the EGA screen is about 14/10; therefore, the default *Windosize* is now -14, 14, -10, 10.
- A new menu entry "o" (oh, not zero) is added to the NUMERICS menu for computing orbits in both forward and backward time. This entry does not work with difference equations.
- The figures in Chapter 4 and 5 are made at the EGA resolution. All other figures are at the CGA resolution.
- With user-defined second-order time-periodic systems, the Poincaré map gives only one point, because the solutions are in $\mathbf{R}^3$, not on a cylinder. For example, with a user-defined equation, it is not possible to reproduce *Figure 7.20*. This will be remedied in the next version.

Acknowledgments

Many students and colleagues contributed a great deal to both the conception and the realization of PHASER. First and foremost, I would like express my heartfelt thanks to my students Lisa Heavey, Matthew Merzbacher, Manijeh Shayegan, Mark Sommer, and Michael Strickman. They spent endless hours, both day and night, writing almost all of the code for PHASER through its many versions. Without their energetic collaborations, PHASER would have forever remained only an idea.

Stepping back in time, my graduate school mentors Al Kelley and Ralph Abraham persuaded me that experimental dynamics could be an enjoyable and indispensable complement to classical mathematics. Later at Brown, the graphical vision and enthusiasm of Andries van Dam, as well as his material support, gave me the courage to blend mathematics with computer science. My adventures into higher dimensions with Thomas Banchoff, Frederic Bisshopp, David Laidlaw, and David Margolis were equally invaluable in sharpening my skills and heightening my appreciation of computer graphics. The Lefschetz Center for Dynamical Systems and the Division of Applied Mathematics of Brown University, by providing the necessary care-free, yet stimulating, working environment, enabled me to toy with nonstandard pursuits. In particular, Wendell Fleming encouraged me to design a new course, and Lawrence Sirovich stopped by every Monday afternoon to ask if the book were done.

On the mathematical side, I have been most fortunate, during the past three years, to share the same hallway and lunch table with Jack Hale. His expert knowledge and friendship have had a profound effect on me, as well as on PHASER. My association with Donald Aronson has likewise been critical in shaping the project. I have, in addition, received mathematical, material, or moral support from Shui-Nee Chow, John Guckenheimer, Philip Holmes, John Mallet-Paret, Donald McClure, Susan Schmidt, Thomas Sharp, and Jorge Sotomayor.

In the final stages of the manuscript, I have also benefitted from the assistance of Philip Davis, Susan McGowen, Wayne Nagata, Brian Weibel, and Andy Young. Above all, however, Lee Zia and Nancy Lawther not only read the entire book, but also made sure that I enjoyed the difficult task of writing and rewriting.

Thank you all; I could not have done it without you.

Contents

Part I: Mathematical Synopsis

Part II: Handbook of PHASER

Part I

Mathematical Synopsis

Chapter 1

What is a Differential Equation ?

The purpose of this chapter is to give an illustrated summary of some of the basic analytical and geometric ideas from ordinary differential equations (see the figures at the end of the chapter). For a more leisurely exposition of this material, you can consult sections 3.1, 4.1, and 4.4 of Braun [1983], for example, or sections 2.3, 7.1, and 9.2 of Boyce & DiPrima [1977]. You may also enjoy reading Chapter 1 of Hirsch & Smale [1974].

1.1. Systems of Differential Equations

Throughout our study of differential equations, from both the geometrical and the numerical points of view, we will consider systems of simultaneous *first-order differential equations* of the following form:

$$\begin{aligned} \frac{dx_1}{dt} &= f_1(x_1, \ldots, x_n) \\ \frac{dx_2}{dt} &= f_2(x_1, \ldots, x_n) \\ &\vdots \\ \frac{dx_n}{dt} &= f_n(x_1, \ldots, x_n) . \end{aligned} \tag{1.1}$$

Here $x_i = x_i(t)$, for $i = 1, \ldots, n$, are n unknown real-valued functions of a real variable t, and dx_i/dt are their derivatives (we will also use x_i' for these derivatives). The f_i are n given functions of the vector variable $\mathbf{x} = (x_1, x_2, \ldots, x_n)$. The variables x_i are called the *dependent variables,* while t is referred to as the *independent variable.* The number of equations n, which is also the number of dependent variables, is referred to as the *dimension* of the system. Notice that the functions f_i do not depend on t explicitly. Such a system is said to be *autonomous* because the derivatives are determined solely by **x**. We will see shortly that not only is the class of differential equations described by (1.1) quite large, but that almost any equation can be put into this form. In fact, before entering your own equations into PHASER, it will be necessary for you to do so.

In addition to the system of differential equations (1.1), we will often be given *initial conditions* for the functions x_i. We will write these in the form

$$x_1(t_0) = x_1^0\,, \quad x_2(t_0) = x_2^0\,, \ldots, \quad x_n(t_0) = x_n^0\,. \tag{1.2}$$

In almost all cases we will take t_0 to be 0. The system of equations (1.1), together with the initial conditions (1.2), is called an *initial-value problem.* Solving an initial value problem consists of finding n functions $x_i(t)$ satisfying both (1.1) and (1.2). For example, the solution of the initial value problem

$$\frac{dx_1}{dt} = x_2\,, \qquad \frac{dx_2}{dt} = -x_1\,, \tag{1.3}$$

$$x_1(0) = 5\,, \qquad x_2(0) = 5 \tag{1.4}$$

is given by the two functions

$$x_1(t) = 5\sqrt{2}\,\sin(\frac{\pi}{4} + t) \quad \text{and} \quad x_2(t) = 5\sqrt{2}\,\cos(\frac{\pi}{4} + t)\,. \tag{1.5}$$

- The system of differential equations (1.3) describes the motion of a *linear harmonic oscillator,* which is a special case of the equation *linear2d* stored in the library of PHASER. We will use these equations repeatedly to illustrate various concepts.

If the functions f_i in (1.1) satisfy certain reasonable conditions, then an initial value problem has a *unique* solution for values of t in an interval containing t_0. For a detailed discussion of basic existence and uniqueness theory, see, for example, Boyce & DiPrima [1977, p. 70], Braun [1983, p. 64], and Hirsch & Smale [1974, p. 159]. Unfortunately, despite the impression one is often given in introductory courses on differential equations, there are no known methods for explicitly solving most initial value problems. While this is certainly very disturbing, luckily it is not necessary to find explicit solutions in most applications;

indeed, carefully obtained partial information can go a long way. There are two successful ways to proceed. First, we can use numerical algorithms to approximate solutions. With the help of computers, this has become a relatively routine task, one which is precisely what PHASER is good at. We need to be very careful in using this approach, however, as it is easy to generate meaningless or misleading information. We will say more about numerical solutions of differential equations in Chapter 2. The second approach falls under the heading "qualitative theory". This is a considerably more difficult subject, but it is possible to make good use of it in applications, without full mastery of the details. In practice, especially when analyzing specific equations, the best thing to do is to use both approaches in tandem. During the past decade this combination has been the source of unparalleled excitement in our subject and, with the help of PHASER, we will try to capture some of this excitement in the following pages.

1.2. Geometrical Interpretation

Let us reconsider the system of differential equations (1.1) and its solutions from a more geometric point of view. For simplicity, we will restrict ourselves to dimension two ($n = 2$). To each point $\mathbf{x}$ in $\mathbf{R}^2$ -- that is, to each vector $\mathbf{x} = (x_1, x_2)$ -- we can associate the vector $\mathbf{f}(\mathbf{x}) = (f_1(\mathbf{x}), f_2(\mathbf{x}))$, which we should think of as being based at $\mathbf{x}$. In other words, we assign to $\mathbf{x}$ the directed line segment from $\mathbf{x}$ to $\mathbf{x} + \mathbf{f}(\mathbf{x})$. In the case of equation (1.3), for example, at the point (5, 5) we picture an arrow pointing from (5, 5) to (5, 5) + (5, –5) = (10, 0). The collection of these arrows is called the *vector field* generated by the first-order autonomous differential equation; see *Figures 1.2* and *1.3*.

Every solution of (1.1) defines a curve $\mathbf{x}(t) = (x_1(t), x_2(t))$ in the plane, parametrized by the independent variable t. If an initial condition is specified, then the solution curve passes though this point when $t = t_0$. For instance, the solution of the linear harmonic oscillator problem (1.3-5) is the circle centered about the origin and passing through the point (5, 5). The right-hand side of (1.1) gives the tangent vector $\mathbf{x}'(t) = (x_1'(t), x_2'(t))$ to the curve. If we interpret the independent variable t as *time*, then the solution curve $\mathbf{x}(t)$ can be thought of as the path of a particle moving in the plane, and the vector field as its velocity vectors; see *Figure 1.4*.

The *graph* of a solution curve is the set of all points $(t, x_1(t), x_2(t))$, and its picture requires a three-dimensional space. For example, in the case of the linear harmonic oscillator, the solution (1.5) is a helix in the (t, x_1, x_2)-space; see *Figure 1.5*. If we project this helix into the (x_1, x_2)-plane, we get the circle described above, and the solution traces this circle with period 2π, as t runs from 0 to infinity.

In the case of first-order autonomous systems, it is not only "safe" but preferable to study the projections of solutions into the (x_1, x_2)-plane, which is called the *phase-plane* (in higher dimensions the *phase-space*). The projections of solutions of (1.1) into the phase-space $(x_1, \ldots, x_n)$ are called *orbits* or *trajectories*, and the collection of all orbits is called the *phase-portrait*. The phase-portrait of the linear harmonic oscillator (1.3) is illustrated in *Figure 1.6*.

Two special types of orbits are very important in the geometric study of phase portraits. An orbit $\mathbf{x}(t)$ is called an *equilibrium point* if it is constant for all values of t; that is, if $\mathbf{x}'(t) = \mathbf{f}(\mathbf{x}) = 0$. For example, the origin $\mathbf{x} = (0, 0)$ is the only equilibrium point of the linear harmonic oscillator (1.3). An orbit $\mathbf{x}(t)$ is called a *periodic orbit* with *period T* if it repeats itself over every time interval of length T; equivalently, if $\mathbf{x}(t + T) = \mathbf{x}(t)$ for all values of t. For instance, all orbits of (1.3) except the origin are periodic orbits with the same period 2π.

1.3. Higher Order Equations

To study a higher-order differential equation using PHASER, and also to take advantage of the geometric theory, it is necessary to convert such an equation into an equivalent first-order system. In this section we describe how to accomplish this.

Consider an nth-order differential equation for a single dependent variable y that can be solved for the highest order derivative of y as an explicit function of the lower order derivatives:

$$\frac{d^n y}{dt^n} = F\left(y, \frac{dy}{dt}, \ldots, \frac{d^{n-1}y}{dt^{n-1}}\right). \tag{1.6}$$

Furthermore, suppose that initial conditions of the form

$$y(t_0), \quad \frac{dy}{dt}(t_0), \ldots, \quad \frac{d^{n-1}y}{dt^{n-1}}(t_0) \tag{1.7}$$

have been specified. Then, by introducing the variables

$$x_1(t) = y, \quad x_2(t) = \frac{dy}{dt}, \ldots, \quad x_n(t) = \frac{d^{n-1}y}{dt^{n-1}}, \tag{1.8}$$

the nth-order equation (1.6) can be converted into the following equivalent first-order system:

$$\frac{dx_1}{dt} = x_2$$

$$\frac{dx_2}{dt} = x_3$$

$$\vdots \tag{1.9}$$

$$\frac{dx_n}{dt} = F(x_1, \ldots, x_n),$$

with the initial conditions

$$x_1(t_0) = y(t_0), \quad x_2(t_0) = \frac{dy}{dt}(t_0), \ldots, \quad x_n(t_0) = \frac{d^{n-1}y}{dt^{n-1}}(t_0).$$

Notice that the system (1.9) is a special case of (1.1).

Example. Convert the initial value problem for the second-order differential equation governing the motion of a linear harmonic oscillator

$$\frac{d^2y}{dt^2} + y = 0, \quad y(0) = 5.0, \quad \frac{dy}{dt}(0) = 5.0$$

into an equivalent system of two first-order differential equations.

Solution. Using substitution (1.8), let $x_1(t) = y$ and $x_2(t) = \frac{dy}{dt}$. Then,

$$x_1' = x_2, \quad x_2' = -x_1, \quad x_1(0) = 5.0, \quad x_2(0) = 5.0.$$

Example. Convert the following third-order differential equation into an equivalent first-order system:

$$\frac{d^3y}{dt^3} = 2y\left(\frac{dy}{dt}\right)^2 + 4y.$$

Solution.

$$x_1' = x_2, \quad x_2' = x_3, \quad x_3' = 2x_1x_2^2 + 4x_1.$$

1.4. Non-autonomous Equations

For computational purposes, a system of non-autonomous differential equations -- that is, a system in which the independent variable t appears explicitly in f_i -- must be converted into an equivalent autonomous system. This can be accomplished by introducing the following additional first-order differential equation, with the indicated initial condition:

$$x_{n+1}' = 1.0 , \quad x_{n+1}(0) = t_0 .$$

The solution of this last equation is $x_{n+1}(t) = t + t_0$. Therefore, we can replace t by x_{n+1} in all f_i on the right-hand side, making the system autonomous at the expense of increasing its dimension.

Example. Convert the following first-order non-autonomous equation (cf. Boyce & DiPrima [1977], p. 360) into a system of autonomous equations:

$$x_1' = 1.0 - t + 4.0x_1 , \quad x_1(0) = 1.0 . \tag{1.10}$$

Solution. This initial value problem is equivalent to the system of two equations below, with the indicated initial values:

$$x_1' = 1.0 - x_2 + 4.0x_1 , \quad x_2' = 1.0 , \tag{1.11}$$

$$x_1(0) = 1.0 , \quad x_2(0) = 0.0 .$$

Notice that the second equation is "decoupled" from the first one, and has the solution $x_2(t) = t$.

Example. Write the following equation of motion of a mass-spring-dashpot system (cf. Boyce & DiPrima [1977], p. 135 and Braun [1983], p. 163) as an autonomous first-order system:

$$m\frac{d^2y}{dt^2} + c\frac{dy}{dt} + ky = F_0 \cos(at) . \tag{1.12}$$

Here m, c, k, F_0 and a are given constants determined by characteristics of the mass, dashpot, spring, driving force, and driving frequency, respectively.

Solution. This is a second-order non-autonomous equation, and it is equivalent to the system of three first-order equations below:

$$x_1' = x_2 ,$$

$$x_2' = -\frac{c}{m}x_2 - \frac{k}{m}x_1 + \frac{F_0}{m}\cos(ax_3) , \tag{1.13}$$

$$x_3' = 1.0 .$$

- The three-dimensional differential equation (1.13) is stored in the library of PHASER under the name *vibration.*

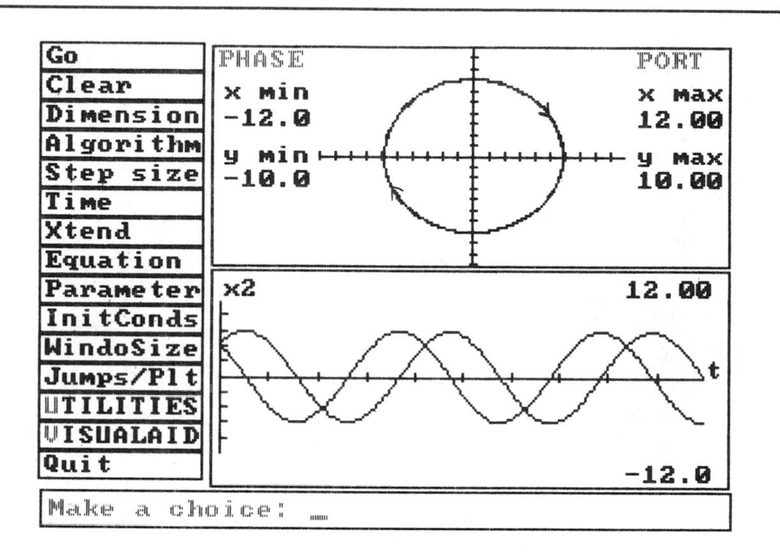

Figure 1.1. Plot of $x_1(t)$ vs. $x_2(t)$ in the top view, and graphs of $x_1(t)$ and $x_2(t)$ vs. t in the bottom view of the functions (1.5) for t from 0 to 15. This is a screen image of PHASER. The function of the "menu" on the left will be explained in Chapter 4.

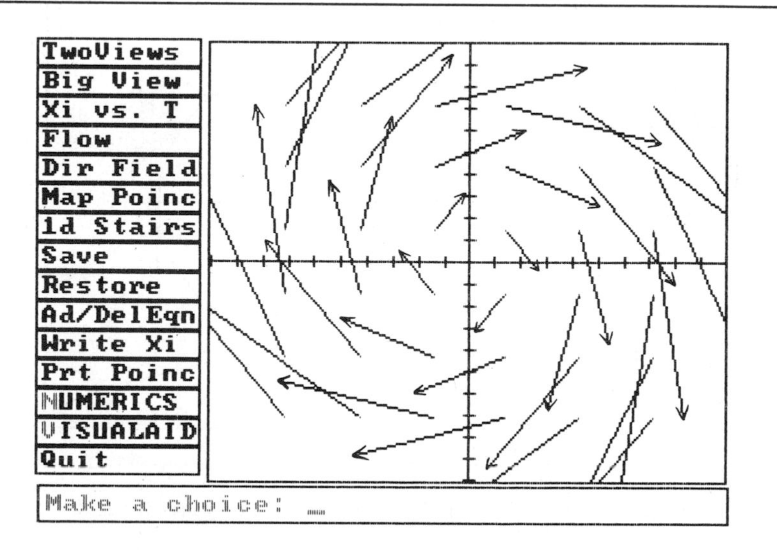

Figure 1.2. Vector field generated by the differential equations of linear harmonic oscillator (1.3) on the (x_1, x_2)-phase plane.

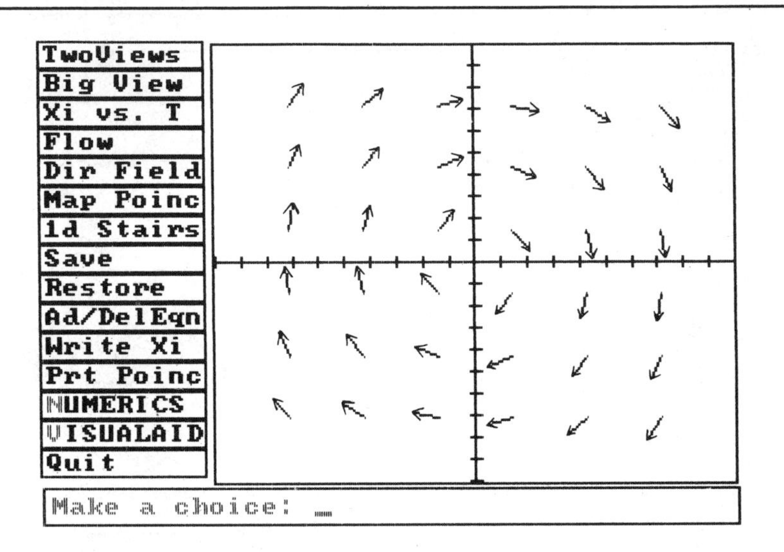

Figure 1.3. "Direction field" (that is, the vector field where all vectors are normalized to have an equal "small" length) of linear harmonic oscillator (1.3).

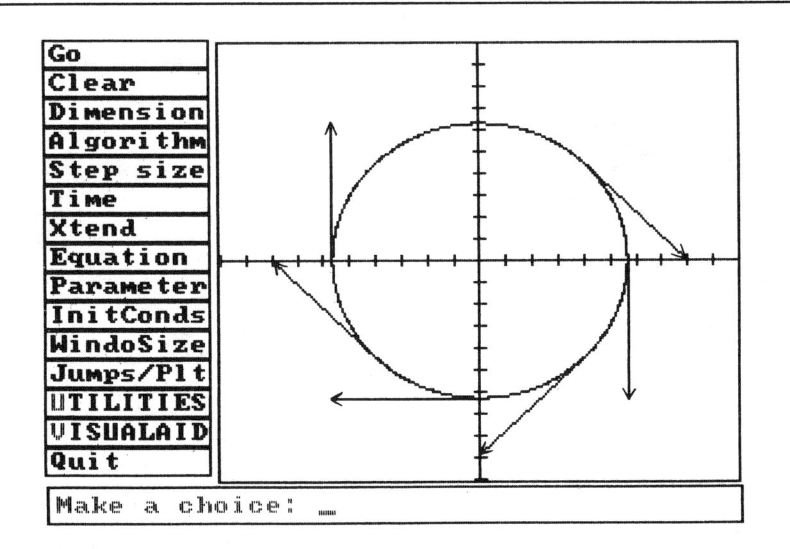

Figure 1.4. The orbit (1.5) in (x_1, x_2)-plane and its tangent vectors.

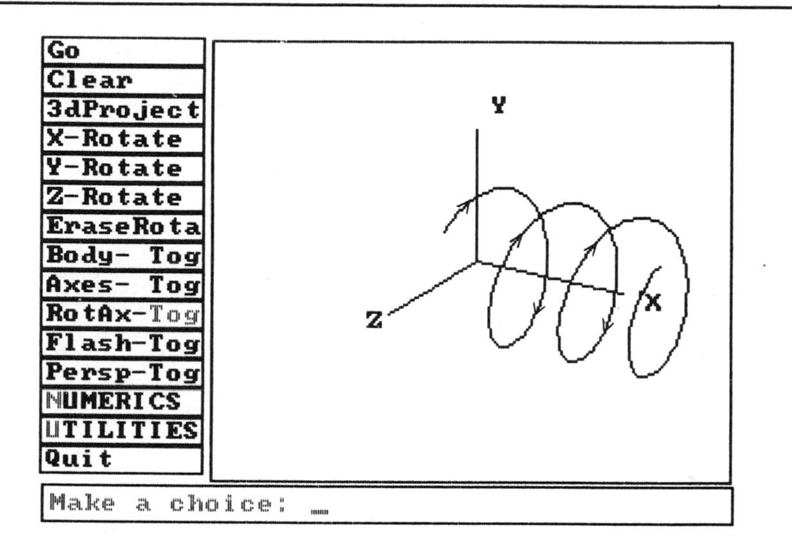

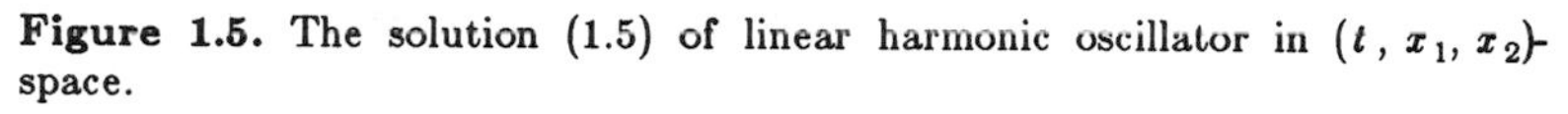

Figure 1.5. The solution (1.5) of linear harmonic oscillator in (t, x_1, x_2)-space.

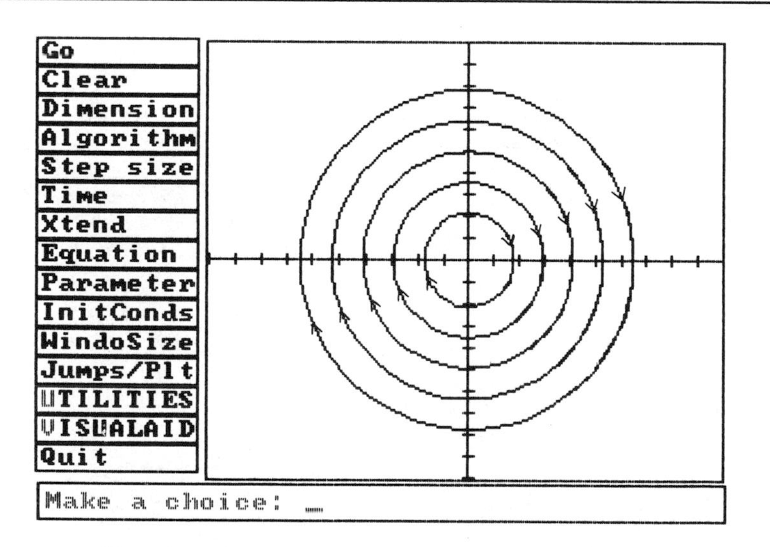

Figure 1.6. Phase portrait of linear harmonic oscillator.

Chapter 2

Numerical Methods

As we have already mentioned, most initial value problems do not have "closed-form" solutions. To obtain approximate solutions, especially in applications, one must resort to numerical methods. This, in fact, is how PHASER generates the orbits you will see in the illustrations. In this chapter, we will briefly discuss what it means to solve an initial value problem (1.1-2) using numerical algorithms, and also give some practical guidelines. For a good elementary introduction to this subject, you should start with Chapter 8 of Boyce & DiPrima [1977], which contains a discussion of the algorithms used by PHASER. If you wish, you can follow this up with more advanced books such as Conte & deBoor [1972] and Gear [1971].

2.1. Overview of Numerics

Computers can manipulate only finite sets of numbers, but they are extremely good at performing well-defined algorithmic tasks repeatedly. Guided by these two facts, an initial value problem can be "solved" on the computer as follows.

We first specify a final value t_{end} of the independent variable t, up to which a solution curve is to be computed. Then we select a finite number of points $[t_0, t_1, \ldots, t_k, \ldots, t_{end}]$ belonging to the time interval $[t_0, t_{end}]$. For simplicity, we will require that these points be equally spaced. The distance $h \equiv t_{k+1} - t_k$ between two consecutive points is called the ***step size***. Next, we calculate approximate values of the solution $\mathbf{x}(t)$ at these equally spaced points, much like the tabulated values of trigonometric and logarithmic functions found in older calculus books. By connecting the consecutive values of $\mathbf{x}(t)$ with straight lines, we can determine a "piecewise linear" approximation to the solution on the entire interval $[t_0, t_{end}]$. If the step size is sufficiently small, this procedure yields a good approximate picture of the solution curve.

Now, the only thing we know about a solution $\mathbf{x}(t) = (x_1(t), \ldots, x_n(t))$ is that it satisfies a given differential equation, and that its value at $t = t_0$ is provided by the initial condition $\mathbf{x}(t_0) = (x_1(t_0), \ldots, x_n(t_0))$. Our main task is to find an algorithm to approximate $\mathbf{x}(t_1)$ using this information. Then, knowing $\mathbf{x}(t_1)$, we determine $\mathbf{x}(t_2)$, and so on with the same algorithm.

The simplest such algorithm, attributed to Euler, is given by the following recursive formula:

$$\text{for} \quad i = 1, 2, \ldots, n\,, \qquad \textit{calculate}$$

$$x_i(t_{k+1}) = x_i(t_k) + hf_i(x_i(t_k))\,. \tag{2.1}$$

It is based on the idea of approximating a function by its tangent lines, and geometrically it amounts to moving a small distance, scaled by the step size h, along the vector field of a differential equation (see section 1.2).

In Euler's algorithm (2.1) we are free to choose the step size h, or, equivalently, the number of points in the interval $[t_0, t_{end}]$. Unfortunately, the "best" choice of h is difficult to determine in practice. It depends on the problem, the computer used, and the programming style. In theory, however, accuracy of approximations is expected to improve with decreasing step size. For Euler's algorithm, this good behavior is illustrated in *Figure 2.1* on the initial value problem (1.3-4) for the linear harmonic oscillator; see also *Figures 2.2-3.*

2.2. A Comparison of Algorithms

In addition to *Euler*, PHASER provides two other algorithms: *Improved Euler* and fourth-order *Runge-Kutta.* Although these are based on more complicated and, in general, more accurate approximation schemes, in usage they are identical to *Euler.* For a detailed discussion of these and many other algorithms, you should consult the references.

To compare the algorithms of Euler and Runge-Kutta, once again consider our central example (1.3), the linear harmonic oscillator. The numerical approximations to the solution (1.5) using *Euler* and *Runge-Kutta* with the same step size $h = 0.1$ are illustrated in *Figures 2.2* and *2.3,* respectively. The discrepancy between the two results is quite striking, especially in the phase plane: *Runge-Kutta* provides an accurate picture of the circular (periodic) orbit; *Euler*, on the other hand, is not only less accurate, but also gives a qualitatively incorrect answer. Instead of staying near the actual periodic orbit, the numerically calculated orbit spirals away to infinity. The qualitative difference between these two pictures should be taken as a serious warning that one must be very careful when solving initial value problems numerically.

2.3. Practical Guidelines

When using numerical methods, careful preparation and experimentation are often necessary to obtain meaningful results. For good practice, you should keep the following guidelines in mind:

- Convert your equations into autonomous first-order systems. All the equations in the permanent library of PHASER are in this form.

- By appealing to some version of the existence and uniqueness theorem, verify first that your initial value problem has a unique solution, so that you will not be looking for a calf under a bull.

- Solutions may not be valid for all values of t, and they may also become unbounded or run into a discontinuity in finite time. In general, it is difficult to determine how large t_{end} can be just by looking at an equation; do your best. For example, consider the initial value problem
$$x_1' = x_1^2 , \quad x_1(0) = 1.0 .$$
It can be readily verified by direct substitution that
$$x_1(t) = \frac{1.0}{1.0 - t}$$
is the solution. Observe that the solution becomes unbounded as $t \to 1.0$.

- If possible, locate special orbits, like equilibria, and determine their stability types (source, sink, saddle, etc.). Especially for two-dimensional problems, this can reveal much about the possible behavior of other orbits.

- There is no universal algorithm that works well for all problems. The choice should be dictated by the problem at hand. If you need to approximate an orbit for a short period of time, do not discount simple algorithms since it is easier to analyze the possible sources of difficulties associated with such algorithms.

- Start computing with a simple algorithm like *Euler*, and a "reasonable" step size h. Then repeat the computations with a step size $h/2$ and compare the results. If the values of $\mathbf{x}(t)$ computed with the two different step sizes differ by an amount larger than you are willing to accept, halve the step size again. Repeat this process until satisfied.

- Theoretically, the algorithms used in PHASER are supposed to yield more accurate results with decreasing step size. In practice, however, this need not be the case; with a smaller step size, more arithmetical operations are required to compute an orbit for the same value of t_{end}. Since computer arithmetic is done in finite precision, this increases the possibility of contaminating computations due to *roundoff errors.* Consequently, once the step size has been halved several times, a simple algorithm may yield unacceptable results. In this case, you should switch to a more sophisticated algorithm like *Runge-Kutta,* and use a bigger step size.

- In those instances where several algorithms with various step sizes give vastly different results for the same problem, turn the machine off for awhile, and analyze the problem further using "classical" mathematics.

```
EQUATION = linear2d        ALGORITHM = Euler        STEP-SIZE = 0.050000
PARAMETERS:
a = 0.000000  b = 1.000000  c = -1.000000  d = 0.000000


     TIME = 14.900000
IC 1: x1 = 0.3585751988    x2 = -10.2516156318

     TIME = 15.000000
IC 1: x1 = -0.6674828024    x2 = -10.2618441126
```

```
EQUATION = linear2d        ALGORITHM = Euler        STEP-SIZE = 0.010000
PARAMETERS:
a = 0.000000  b = 1.000000  c = -1.000000  d = 0.000000


     TIME = 14.900000
IC 1: x1 = 0.1756677039    x2 = -7.6159279144

     TIME = 15.000000
IC 1: x1 = -0.5858015039    x2 = -7.5992179254
```

```
EQUATION = linear2d        ALGORITHM = Euler        STEP-SIZE = 0.001000
PARAMETERS:
a = 0.000000  b = 1.000000  c = -1.000000  d = 0.000000


     TIME = 14.900000
IC 1: x1 = 0.1607742717    x2 = -7.1221295393

     TIME = 15.000000
IC 1: x1 = -0.5510827723    x2 = -7.1029543591
```

Figure 2.1. Last several values of approximations to the initial value problem (1.3-4), using Euler's algorithm with step sizes 0.05, 0.01, and 0.001. We will explain the meanings of "Parameters" and "IC 1:" in Chapter 4.

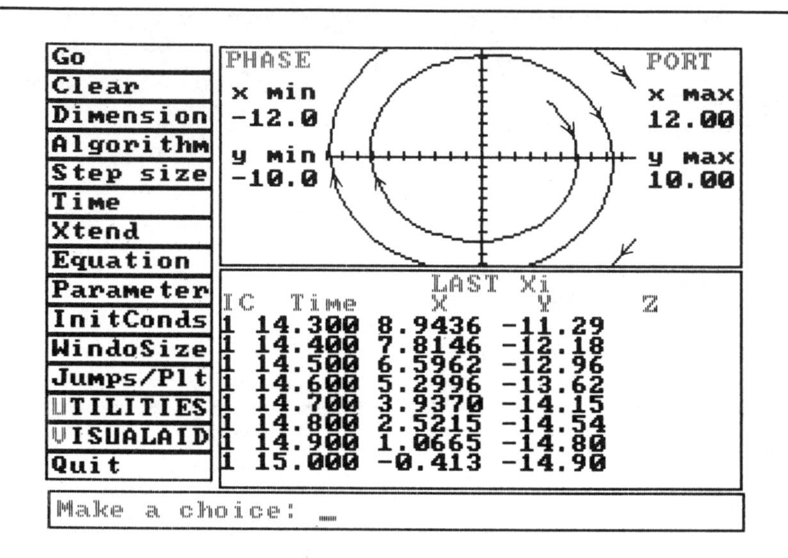

Figure 2.2. Tabulated values and plot of the orbit (1.5) of linear harmonic oscillator, using Euler's algorithm with step size 0.1.

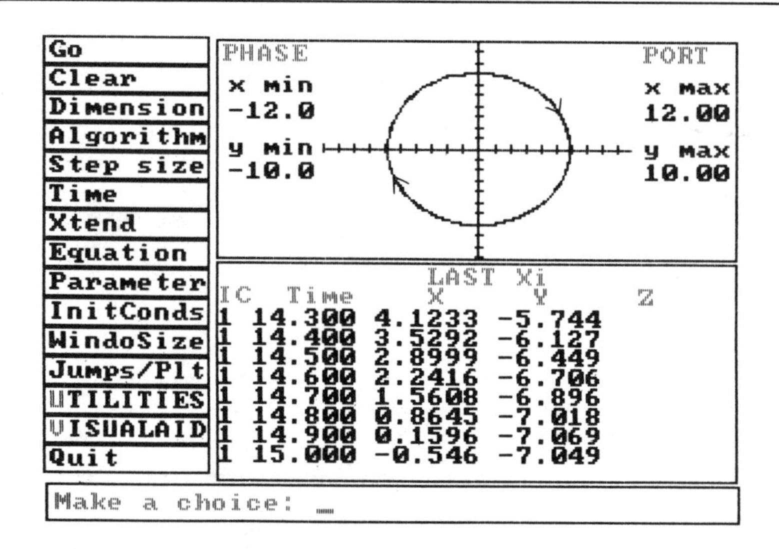

Figure 2.3. Tabulated values and plot of the orbit (1.5) of linear harmonic oscillator, using Runge-Kutta's algorithm with step size 0.1.

Chapter 3

What is a Difference Equation ?

The theory of difference equations, despite its absence from the undergraduate curriculum, is an old and beautiful part of mathematics, one with diverse applications to many subjects: biology, economics, numerical analysis, etc. In a difference equation, change takes place in discrete time intervals. For example, in modeling populations of seasonally breeding animals, it is preferable to use difference equations rather than differential equations because the size of the next generation is largely determined by that of the current one.

Those working in ordinary differential equations are certainly no strangers to a class of difference equations called *discrete dynamical systems.* They have been studying such objects as cross sections to solutions of differential equations since the days of Poincare (see section 3.4). Recently, however, as a consequence of several innocuous looking examples, the subject has been revitalized. The level of activity in this area has become explosive, and the reason behind this late blooming is best summarized by Lanford [1980]:

> The methods used to study (real) smooth transformations of intervals are by and large, elementary, and the theory could have been developed long ago if anyone had suspected that there was anything worth studying. In actual fact, the main phenomena were discovered through numerical experimentation, and the theory has been developed to account for observations. In this respect, computers have played a crucial role in its development.

This chapter is a brief introduction to some of the geometric ideas from the theory of difference equations, also called *discrete dynamical systems* or *maps.* The one-dimensional case is illustrated with two examples, using stair step diagrams. The logistic equation, which has been largely responsible for recent excitement in this area, is the first example. The second is the numerical analysis problem of calculating the square root of a number. Brief descriptions of systems of first-order difference equations and of the reduction of higher-order equations are also included. We conclude the chapter with a short glimpse of planar Poincare maps.

For a good elementary introduction to the subject, you should read the influential, if a bit dated, review article by May [1976] and the book by Devaney [1985]. Depending on your background, you can follow these up with excellent survey articles by Lanford [1980], Nitecki [1981], Whitley [1983], and the book by Collet & Eckmann [1980].

3.1. One-Dimensional Difference Equations

In this section, we discuss first-order difference equations of a single variable. Besides being a natural starting point, these equations possess an extraordinarily rich behavior, and currently form an active area of research.

Let $x(t)$ be an unknown function of a real variable t, and f be a given function of x. A *first-order difference equation* is a relationship of the form

$$x(t+1) = f(x(t))$$

between the successive values of x when t is a nonnegative integer. We will prefer to write the equation above as

$$x^{k+1} = f(x^k) \qquad \text{for} \quad k = 0, 1, 2, 3, \cdots . \tag{3.1}$$

In addition to (3.1), we will often be given an initial condition of the form

$$x(0) = x^0 . \tag{3.2}$$

Equation (3.1), together with (3.2), is called an initial-value problem for a first-order difference equation. Solving an initial-value problem consists of finding a sequence of points

$$x^0, \ x^1, \ x^2, \ x^3, \ \cdots$$

satisfying (3.1) and (3.2). The elements of this sequence are also called the *iterates of* x^0. For example, the solution of the initial value problem

$$x^{k+1} = 0.5x^k , \qquad x^0 = 5.0 \tag{3.3}$$

is given by the following sequence of numbers:

$$5.0, \quad 2.5, \quad 1.25, \quad 0.625, \quad 0.3125, \quad \cdots .$$

Notation: We will use the notation $f^{(k)}$ to denote the composition of the function f with itself k times, and will call it the kth *iterate* or the kth *power* of f. Consequently, the kth iterate of an initial point x^0 will be denoted by $f^{(k)}(x^0)$.

It is important to keep in mind that in difference equations, unless otherwise stated, we will use superscripts to denote the iterates of a number, not its powers. Notice also that an orbit of a first-order difference equation is a set of discrete points, not a continuous curve.

As with differential equations, most difference equations do not have closed-form solutions. However, solving a difference equation on the computer is quite easy. Given a function $f(x)$ and a starting point x^0, we evaluate $f(x^0)$, put the result back into the function, and so on. There are no numerical algorithms or step sizes to worry about. The only possible source of error is roundoff error due to finite precision arithmetic on the computer. Still, this can be a serious problem with no easy resolution even on simple-looking equations, as illustrated by Curry [1979]:

> In order to demonstrate the effects of computational error on the iteration of Henon's transformation (a quadratic difference equation in two variables, stored in the library under the name *henon*) consider the following simple experiment: Given an initial condition, compute the sixtieth iterate of the transformation using two different machines (a CDC 7600 and a CRAY-1) - both machines carry fourteen significant digits in single precision. We found that there was no agreement in the output of the machines by the sixtieth iterate of the transformation!

Aside from the numerical difficulty above, for some equations it may not even be possible to calculate successive values of x^k. For example, consider the following first-order difference equation:

$$x^{k+1} = -\sqrt{x^k}\ .$$

The function $f(x) = -\sqrt{x}$ is defined only for $x \geq 0$. Starting with any $x^0 > 0$, we get $x^1 < 0$; hence we cannot calculate x^2.

We now turn to the geometric method of ***stair step diagrams*** for following orbits of one-dimensional difference equations. Hopefully, this will alleviate some of the pessimism cast by the remarks above. Since $x^{k+1} = f(x^k)$, we first draw the graph of the function f in the (x^k, x^{k+1})-plane. Then, given x^0, we find x^1 by drawing the vertical line through x^0. This line intersects the graph of the function at the point (x^0, x^1) because $f(x^0) = x^1$. Next, from the point (x^0, x^1) draw the horizontal line. This intersects the 45° line at the point (x^1, x^1). From this point draw the vertical line. This intersects the graph of the function at the point (x^1, x^2) because $f(x^1) = x^2$. Continue the process to find x^3, and so on. In *Figure 3.1* we have carried out the first several iterations of the initial value problem (3.3).

- The one-dimensional difference equation (3.3) is stored in the library under the name *dislin1d.* For further information on the graphical facilities of PHASER for plotting stair step diagrams, you should consult the *1d Stairs* entry in sections 6.2 and 6.4, as well as Lesson 11.

A good place to start the study of a difference equation, both in theory and in applications, is by finding its fixed points or periodic orbits. A point x^* is called a *fixed point* of equation (3.1) if

$$x^* = f(x^*) . \tag{3.4}$$

In other words, x^* is a constant solution. Geometrically, a fixed point is where the graph of f intersects the 45° line, because $x^{k+1} = x^* = x^k$. For example, $x^* = 0$ is the only fixed point of equation (3.3).

A cycle of p distinct points

$$x^k = f^{(k)}(x^0) \qquad \text{for } k = 0, 1, \ldots, p-1, \quad \textit{with} \quad f^{(p)}(x^0) = x^0$$

is called a ***periodic orbit*** of period p. Notice that a periodic orbit of period p of the equation (3.1) corresponds to p distinct fixed points of the difference equation

$$x^{k+1} = f^{(p)}(x^k) . \tag{3.5}$$

Therefore, periodic orbits of period p can be found geometrically by intersecting the graph of $f^{(p)}$ with the 45° line. Observe that a fixed point is a periodic orbit of period 1.

Fixed points and periodic orbits can also be found analytically by solving (3.4) and (3.5), respectively. This is usually a formidable task. Certain fixed points and periodic orbits, however, can be located numerically. A periodic orbit is called *asymptotically stable* if all orbits starting "near" it converge to the periodic orbit as $k \to \infty$. To find an asymptotically stable periodic orbit, we need to know only its approximate location because the orbit of any initial condition near the periodic orbit eventually converges to that periodic orbit. This is an important technique in numerical analysis; see section 3.2.

The Logistic Equation: Now, we will illustrate some of the concepts introduced above using a specific first-order difference equation. Let x^k be the size of the kth generation of a population and a be a parameter reflecting intrinsic growth rate. Then, one of the simplest models describing the growth of the population is given by

$$x^{k+1} = ax^k (1.0 - x^k) ; \tag{3.6}$$

see, for example, Maynard Smith [1968] and May [1976]. Equation (3.6) is referred to as the *logistic equation*, and it has been a major reason for the recent surge of activities in difference equations. Here is a glimpse of the fascinating properties of (3.6). For the parameter value $a = 2.7$, the logistic equation has two fixed points at $x^* = 0$ and $x^* = 1 - 1/a \approx 0.62962...$. The first point is unstable; that is, some of the orbits starting near the fixed point may run away from it. The second, however, is asymptotically stable, as shown in *Figure 3.2.* For the parameter value $a = 3.41$, equation (3.6) still has two fixed points: $x^* = 0.0$ and $x^* \approx 0.70674...$, but neither one is asymptotically stable. In fact, almost any initial condition converges to the periodic orbit $\{0.5380... , 0.7804...\}$ of period 2. The graph of the second iterate of the logistic equation has four fixed points: two are unstable, but the other two correspond to the stable periodic orbit above; see *Figure 3.4.* As the parameter a is increased further, various periodic orbits of different periods appear and disappear. There are also nonperiodic orbits: at $a = 4.0$, a single orbit fills up almost the entire unit interval in a seemingly random fashion. The study of changes in a differential or difference equation as parameters are varied is called *bifurcation theory.* This is currently one of the most active areas of mathematical research, and we will investigate many of the important bifurcations in more detail later.

- The one-dimensional difference equation (3.6) is stored in the library of PHASER under the name *logistic.* For more illustrations, see Lesson 11 in Chapter 5.

3.2. Application: Calculating $\sqrt{2}$

A common theme in numerical analysis is to convert problems to difference equations, because computers are well suited for solving such equations. In fact, many algorithms for finding zeros of functions, integrating differential equations, optimization, etc. are difference equations in disguise. In this section we cast the problem of calculating $\sqrt{2}$ as one of locating an asymptotically stable fixed point of a certain difference equation.

The equation $x^2 = a$ (the square of x, not the second iterate) for $a > 0$ can be written in the form:

$$x = \frac{1}{2}(x + \frac{a}{x}) .$$

Therefore, $\sqrt{a}$ is a fixed point of the following first-order difference equation:

$$x^{k+1} = \frac{1}{2}(x^k + \frac{a}{x^k}) . \qquad (3.7)$$

This equation has two fixed points at $x = \sqrt{a}$, $x = -\sqrt{a}$, and they are both asymptotically stable. Positive initial conditions converge to $\sqrt{a}$, while negative initial conditions converge to $-\sqrt{a}$. For example, for $a = 2.0$ and the starting value $x^0 = 3.0$, we get the orbit

3.00000, 1.83333, 1.46212, 1.41499, 1.41421, 1.41421, $\cdots$.

- This example is a special case of a general scheme, called *Newton-Raphson's method*, for finding zeros of functions (cf. Conte & de Boor [1980]). The one-dimensional difference equation (3.7) is stored in the library of PHASER under the name *newton*; see *Figure 3.5*.

3.3. Systems of Difference Equations

In certain populations, the size of a generation depends on two previous generations. One of the simplest models for the growth of such a population is the following difference equation:

$$N^{k+1} = aN^k(1.0 - N^{k-1}), \tag{3.8}$$

where N^k is the size of the kth generation and a is a parameter reflecting the intrinsic growth rate; see, for example, Maynard Smith [1968, p. 28]. Note that this is the logistic equation (3.6) except that the nonlinear term on the right-hand side contains a time delay of one generation. Consequently (3.8) is known as the *delayed logistic* model.

Equation (3.8) is a *second-order* difference equation because the right-hand side is a function of the *two* previous generations (iterates). For geometrical and computational reasons, it is preferable to transform the second-order equation (3.8) into a pair of simultaneous first-order equations by the introduction of the new variables

$$x_1^k = N^{k-1}, \quad x_2^k = N^k .$$

Then, the delayed logistic equation (3.8) is equivalent to the following system:

$$\begin{aligned} x_1^{k+1} &= x_2^k \\ x_2^{k+1} &= ax_2^k(1.0 - x_1^k) . \end{aligned} \tag{3.9}$$

- A more general form of the two-dimensional difference equation (3.9) is stored in the library of PHASER under the name *dellogis*. This system has fascinating but very subtle bifurcation behavior as the parameter a is varied; see *Figure 3.6* and

Aronson et al. [1982].

Much as in the case of differential equations, it is often necessary, both in theory and in applications, to study general systems of simultaneous first-order difference equations of the form

$$
\begin{aligned}
x_1^{k+1} &= f_1(x_1^k, \ldots, x_n^k) \\
x_2^{k+1} &= f_2(x_1^k, \ldots, x_n^k) \\
&\vdots \\
x_n^{k+1} &= f_n(x_1^k, \ldots, x_n^k),
\end{aligned}
\tag{3.10}
$$

of which (3.9) is a special case. Notice that the functions f_i depend only on the kth iterate of the variables x_i; hence the name first-order. Many of the concepts introduced in section 3.1 for a single first-order difference equation can be generalized to the system (3.10). However, we will refrain from doing so here.

3.4. Planar Poincare Maps

The basic tool used in the study of the asymptotic behavior of recurrent orbits of ordinary differential equations is the Poincare map or local cross section. This is an effective theoretical, as well as experimental, method for dimensions three and higher. In this section we present a brief discussion of this important technique. It may be skipped by beginners. For further information see, for example, Guckenheimer & Holmes [1983, p. 22].

For simplicity, let us consider the case of three dimensions. We select in this space a plane, given by $Ax + By + Cz + D = 0$, and a direction. Then we look at the successive points of intersection of an orbit of a differential equation with this plane as the orbit passes through the plane from one designated side to the other. If the orbit is recurrent -- that is, if it comes back repeatedly to a region of three-space -- then there will be a sequence of intersection points $\mathbf{x}^0$, $\mathbf{x}^1$, $\mathbf{x}^2$, ... , on the plane. The two-dimensional difference equation

$$\mathbf{x}^{k+1} = \mathbf{f}(\mathbf{x}^k),$$

whose solutions are the sequences of such points of intersection, is called the *Poincare map*. The function $\mathbf{f}$, which maps the plane into itself, has the additional property that it is a *diffeomorphism* (both $\mathbf{f}$ and its inverse are differentiable functions). The main importance of Poincare maps stems from the fact that they reflect many essential properties of orbits of differential equations. For example, a simple periodic orbit of a differential equation corresponds to a fixed point of the Poincare map, and the stability type of the orbit can be determined from that of the fixed point. Furthermore, bifurcations of fixed points of Poincare maps can be interpreted as bifurcations of periodic orbits of the corresponding differential equations.

Unfortunately, finding an explicit formula for a Poincare map requires a knowledge of closed-form solutions of the original differential equation. In practice, however, it is quite easy to determine the points of intersection of a numerically computed solution of a differential equation with a given plane. Indeed, we will see that PHASER provides graphical facilities for computing Poincare maps numerically. For example, ***Figures 3.7-8*** show two different orbits of a pair of linear harmonic oscillators and their points of intersection with the horizontal plane through the origin as these orbits cross the plane from the positive side to the negative.

- The equations of a pair of harmonic oscillators are stored in the library of PHASER under the name *harmoscil*. For more information on the example above, see Lessons 12 and 13 in Chapter 5, and the *MapPoinc* and *CutPlane* entries in Chapter 6.

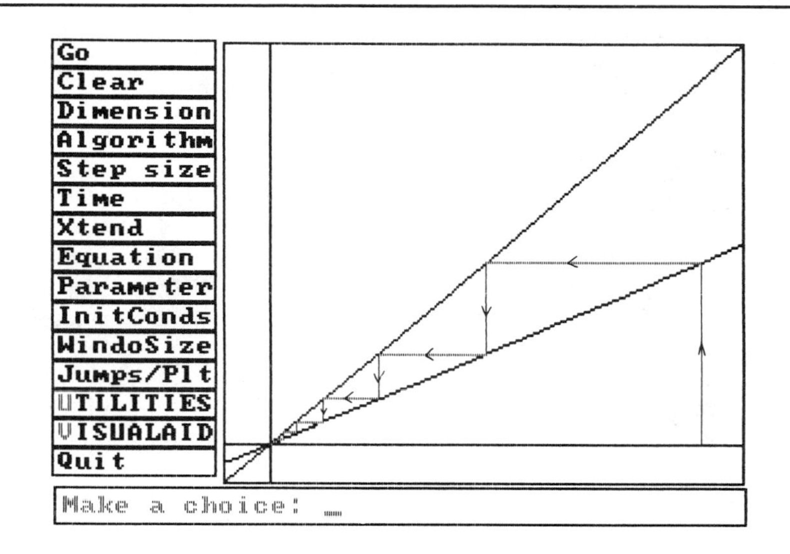

Figure 3.1. The stair step diagram for the first several iterates of the initial value problem (3.3). Notice that the origin is an asymptotically stable fixed point.

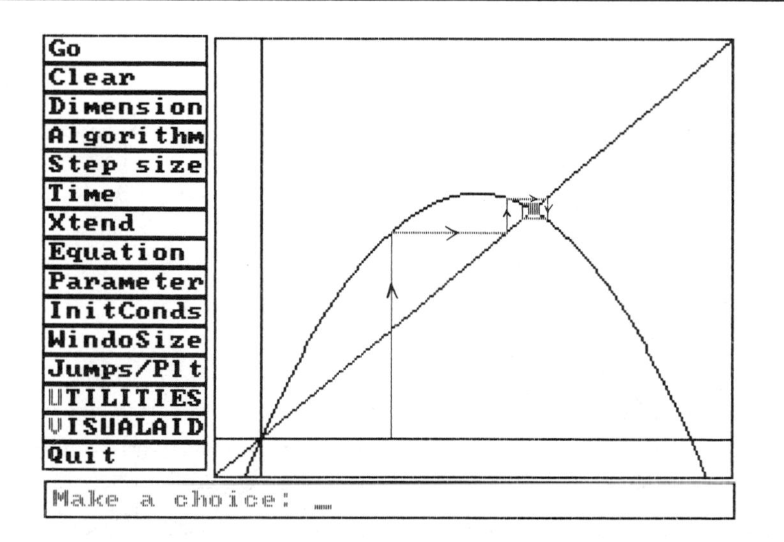

Figure 3.2. The graph of the logistic equation (3.6) for the parameter value $a = 2.7$, and the stable fixed point.

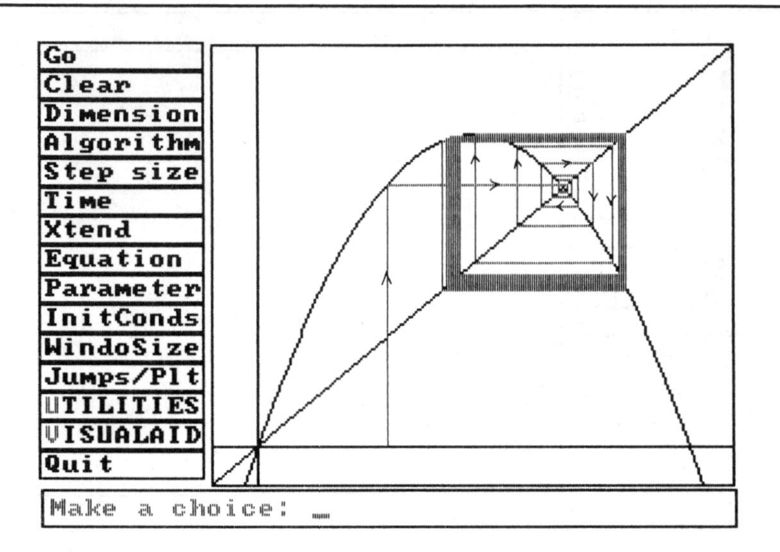

Figure 3.3. The graph of the logistic equation (3.6) for the parameter value $a = 3.41$, and the asymptotically stable periodic orbit of period 2.

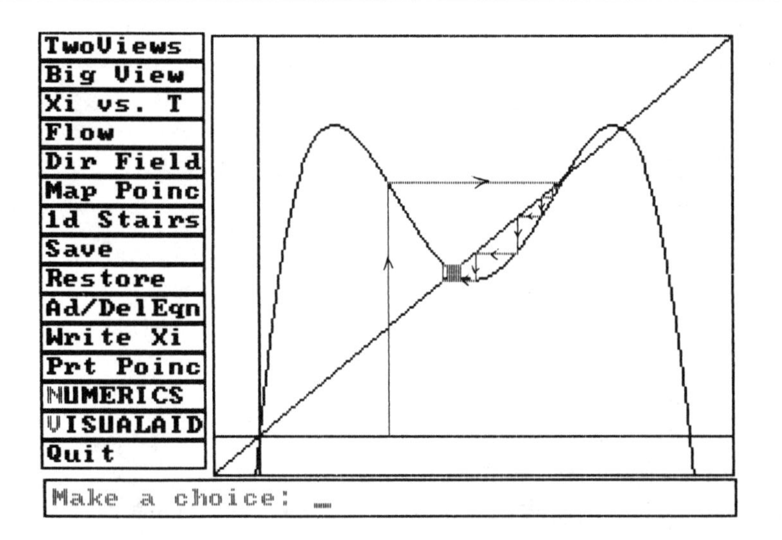

Figure 3.4. The graph of the second iterate, $f^{(2)}$, of the logistic equation (3.6) for the parameter value $a = 3.41$.

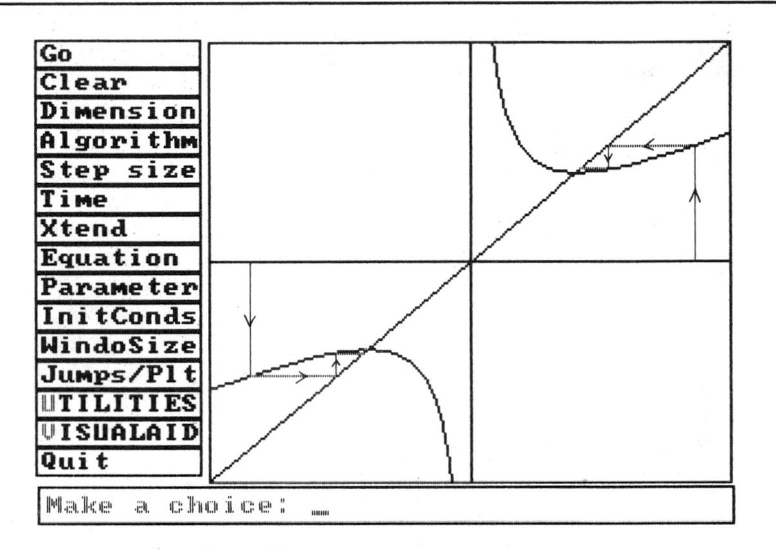

Figure 3.5. Calculating $\sqrt{2}$: stair step diagrams for the two asymptotically stable fixed points of equation (3.7) for the parameter value $a = 2$.

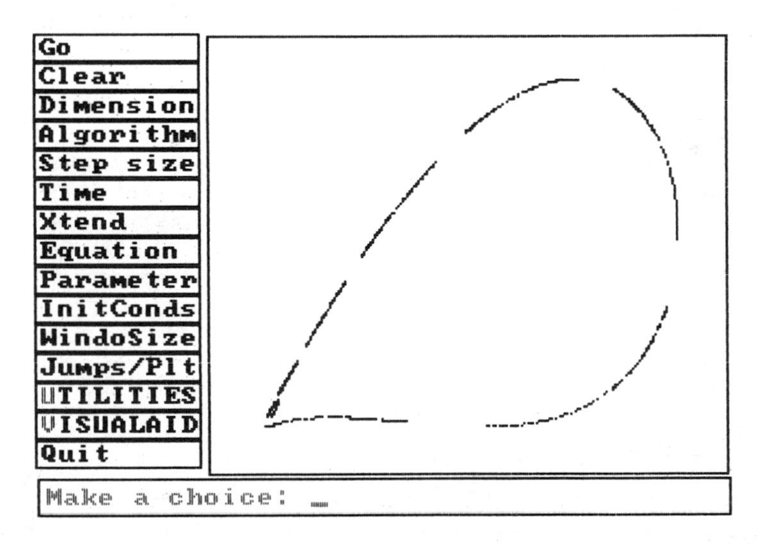

Figure 3.6. The visible attractor of the delayed logistic equation (3.10) with $a = 2.2521185$, consisting of iterates 1000-2500 of the initial point (0.1, 0.2).

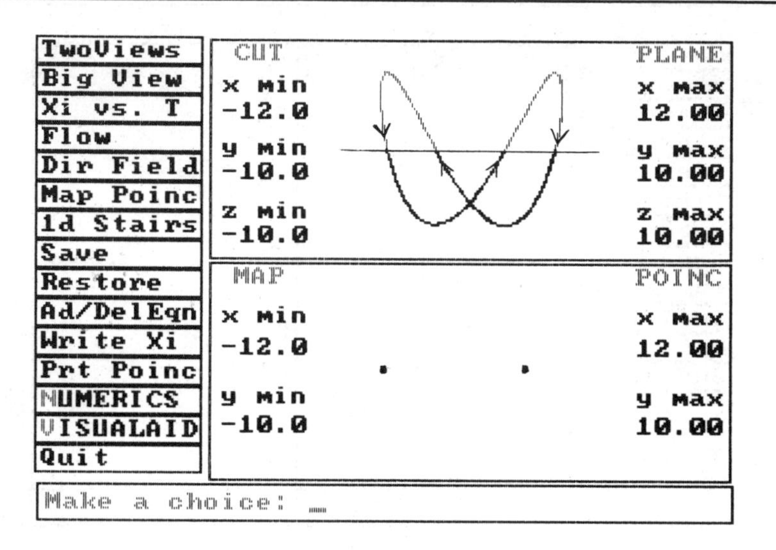

Figure 3.7. A periodic orbit and its Poincare map as the orbit crosses the horizontal plane from the top side of the plane to the bottom (the plane is rotated 90° to look straight on).

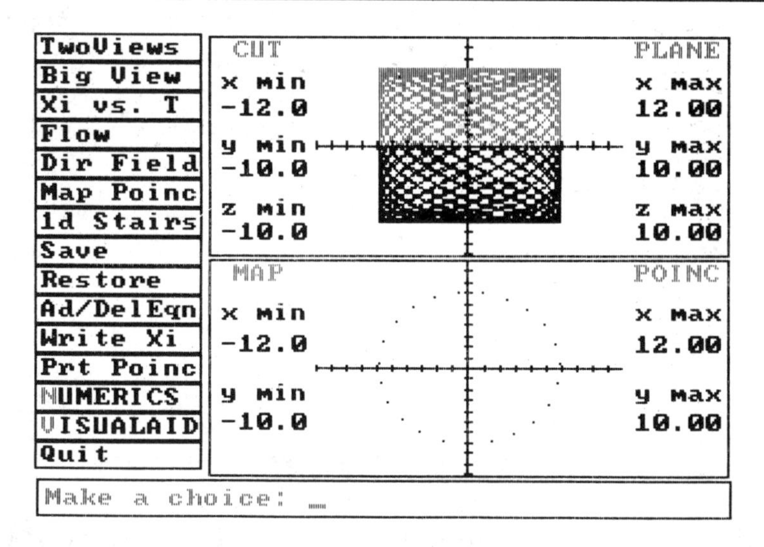

Figure 3.8. A solution curve of a pair of harmonic oscillators (a Lissajous figure), and its Poincare map on the horizontal plane through the origin (the plane is rotated 90° to look straight on).

Part II

Handbook of PHASER

Phase space - the final frontier. These are the voyages of the diskette PHASER, her user's mission to explore strange new attractors, to seek out new equations and new dynamics, to boldly go where no **P**(oin)**C**(are) has gone before.

Chapter 4
Learning to Use PHASER

The purpose of this chapter is to analyze the startup screen layout of PHASER, as illustrated in *Figure 4.2*. Along the way, we will also point out some of the general features of our animator/simulator for dynamical systems.

4.1. The Startup Screen Layout

When PHASER starts up (we will see how to do this in Chapter 5), a brief logo appears, then the screen becomes like the picture in *Figure 4.2*. Let us first see what different parts of the screen represent. The screen is divided into three functionally independent sections: the menu, the message line, and the viewing area.

Menus: The vertical rectangular region on the left, containing key words such as "Go," "Clear," "Dimension," etc., is used to display the menus. The menus are a convenient way of interacting with PHASER. By choosing the appropriate entries of a menu, you can change the current settings of many things -- the step size, for example -- or perform a common task, like numerically computing orbits of differential equations.

To "make a choice" from the current menu, just type in the first character of the entry; you need not hit the <return> or the <enter> key. For example, to change the current step size, which is 0.1, type "s" for "Step Size". You will be asked to "Enter step size". Now, type in,

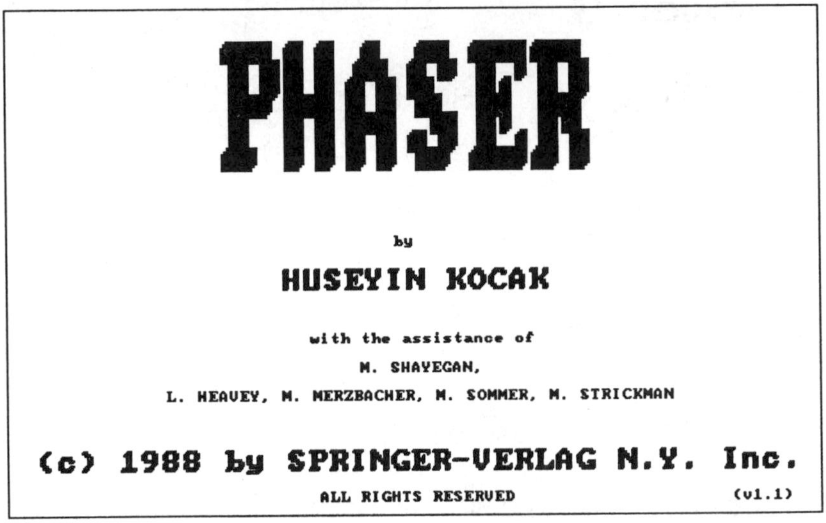
PHASER
by
HUSEYIN KOCAK
with the assistance of
M. SHAYEGAN,
L. HEAVEY, M. MERZBACHER, M. SOMMER, M. STRICKMAN
(c) 1988 by SPRINGER-VERLAG N.Y. Inc.
ALL RIGHTS RESERVED
(v1.1)

Figure 4.1. The logo of PHASER.

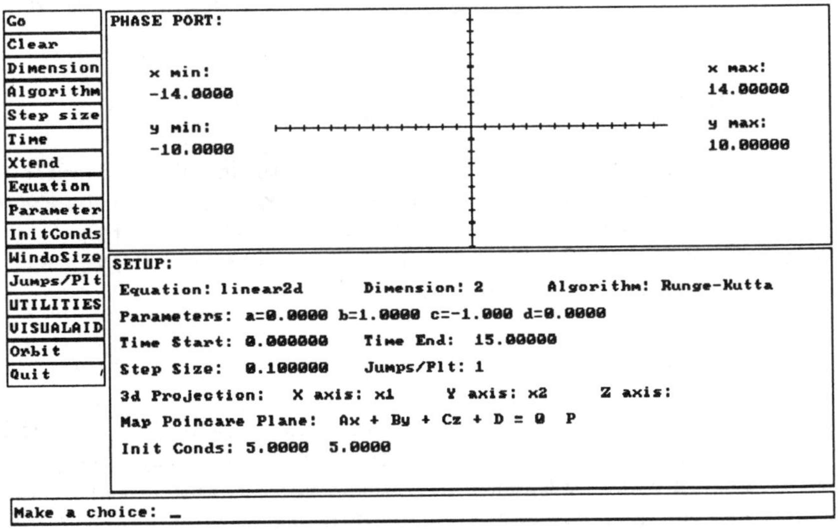
Go
Clear
Dimension
Algorithm
Step size
Time
Xtend
Equation
Parameter
InitConds
WindoSize
Jumps/Plt
UTILITIES
VISUALAID
Orbit
Quit
PHASE PORT:
x min:
-14.0000
y min:
-10.0000
x max:
14.00000
y max:
10.00000
SETUP:
Equation: linear2d
Dimension: 2
Algorithm: Runge-Kutta
Parameters: a=0.0000 b=1.0000 c=-1.000 d=0.0000
Time Start: 0.000000
Time End: 15.00000
Step Size: 0.100000
Jumps/Plt: 1
3d Projection: X axis: x1
Y axis: x2
Z axis:
Map Poincare Plane: Ax + By + Cz + D = 0 P
Init Conds: 5.0000 5.0000
Make a choice: _

Figure 4.2. The startup screen layout of PHASER.

say 0.2, followed by <return>; the new step size will be 0.2, until you change it again. Undoubtedly you are now wondering what might happen if you type "g" for "Go". Please read on; the answer will be revealed shortly.

In addition to the menu seen in *Figure 4.2*, there are many others. We will say more about them in later lessons.

Message Line: This is the thin horizontal rectangle on the bottom portion of the screen, containing the prompt "Make a choice: __ ". The message line is used to display user-supplied information as it is being typed, and, when appropriate, error messages and many other helpful prompts. A growing bar also appears here to monitor the progress of computations; see *Figure 5.1.*

Viewing Area: The portion of the screen containing the two large rectangles on the right is the viewing area. Originally, when you start PHASER, as in *Figure 4.2*, there are two views: the top one is the "Phase portrait" view in which solution curves will be displayed, and the bottom view is "Set up," which contains the current values of computational choices. In addition to "Phase portrait" and "Set up," there are many other views that can be put into the viewing area. We will see how to do this shortly.

Let us now take a closer look at the screen image in *Figure 4.2*, and decipher the information displayed in the two large rectangular views.

Phase Portrait view: This, you recall, is the upper rectangle in which the solution curves will be displayed. The boundaries of the plotting area are given on the sides: the minimum values of the x- and the y-axes are -14.0 and -10.0, and their maximum values 14.0 and 10.0, respectively. If the coordinate values of solutions fall outside this region, computations will continue, but they will *not* be plotted. These boundaries can be changed using the "WindoSize" entry of the menu.

Set Up view: This, again, is the lower rectangle displaying the current values of many things -- step size, initial conditions, etc. -- that we may want to adjust during simulations. The values shown in *Figure 4.2* are the default values which will appear every time PHASER is started up. If you change any of these settings during a session, it will be updated immediately. Here is a line-by-line account of the *Set up* view:

- *Line 1:* "linear2d" is the name of the current equation (4.1) to be studied. It is the general system of two linear ordinary differential equations, given by the following formulas:

$$x_1' = ax_1 + bx_2 \tag{4.1}$$

$$x_2' = cx_1 + dx_2 \,.$$

Here a, b, c, and d are parameters (coefficients) that must be specified. Other equations can be loaded by activating the "Equation" entry of the menu. "2-D" indicates the current dimension, which is 2. It can be changed by activating the "Dimension" entry of the menu. "Runge-Kutta" is the current numerical algorithm to be used in the approximation of solutions. It can be changed by activating the "Algorithm" entry of the menu.

- *Line 2:* The current values of the parameters ($a = 0.0$, $b = 1.0$, $c = -1.0$, $d = 0.0$) of the current equation (4.1) are recorded on this line. They can be changed by activating the "Parameter" entry of the menu.

- *Line 3:* "Start: 0.00" and "End: 15.0" are the boundaries of the time interval during which solutions will be plotted. These bounds can be changed by activating the "Time" entry on the menu.

- *Line 4:* "Step" is the current step size to be used, which is 0.1. "Jumps/Plt" is the number of jumps between plotted points. When this number is set at 1, solutions are plotted at every step. These two numbers can be adjusted by activating the "Step size" and "Jumps/Plt" entries on the menu.

- *Line 5:* This line displays which dependent variables x_i, $i = 1, 2, \ldots, 9$, will be assigned to the three coordinate axes x, y, and z. The information on this line is useful when using a "3d Projection" from a higher dimensional space. We will defer additional details until later lessons. In two dimensions, you can safely forget this line because the variables x_1 and x_2 are always plotted on the x- and y-axes, respectively.

- *Line 6:* This is where the equation of a plane and a negative or positive direction (N/P) are displayed. The plane and the direction are used to take slices and to compute Poincare maps in dimensions greater than two. Once again, we will postpone detailed explanations until later lessons.

- *Line 7:* "IC" stands for initial conditions; the current values are $x_1(0) = 5.0$, $x_2(0) = 5.0$. New initial conditions can be specified by activating the "InitConds" entry on the menu.

And now the mystery of "g" can be resolved: if we type "g" for "Go" to start our simulation, the orbit of the system of linear differential equations

$$x_1' = x_2 \tag{4.2}$$

$$x_2' = -x_1 ,$$

starting at the initial conditions $x_1(0.0) = 5.0$ and $x_2(0.0) = 5.0$, will be computed numerically for the time interval $t = 0.0$ to $t = 15.0$, using the algorithm of Runge-Kutta with step size 0.1 , and will be displayed in the upper rectangle of the viewing area. This, of course, is the initial value problem (1.3-4) for the linear harmonic oscillator discussed in Chapter 1.

4.2. Interacting with PHASER

In preparation for the lessons in Chapter 5, we will now give somewhat more detailed information on how to interact with PHASER using the menus. For brief descriptions of the entries of the menus, see the *PHASER Quick Reference* in Appendix A; full details are given in Chapter 6.

There are three main menus called NUMERICS (the menu in *Figure 4.2*), UTILITIES, and VISUALAID. To "make a choice" from the current menu, type the first character of the name of the entry, in either lower-case or upper-case, but *do not* hit the <return> key. For instance, if you wish to bring up the UTILITIES menu, type "U" or "u". The NUMERICS menu will be erased, and the UTILITIES menu will be drawn in the same place. To bring back the NUMERICS menu, just type "n". To enhance recognition, the first letters of the main menu names are written in green.

If you select a menu entry that requires user input, you will get a prompt in the message line, which shows in parentheses the current value of the expected input. For example, if you choose the "Dimension" entry and the current dimension is 2, you will be prompted with "Enter dimension (2) : __ ". If you want the current dimension to remain 2, just hit <return>. Otherwise, type in a different integer of your choice, followed by <return>. The general rule is that after making a choice from a menu, hit <return> if you wish to retain the current setting of the entry; otherwise, type in your input, followed by <return>.

If you select a menu entry that requires further choices, a submenu appears. If you select the "Algorithm" entry, for instance, a submenu containing four new entries, "Difference," "Euler," "ImpEuler," and

"Runge-Kut," appears; see *Figure 4.3.* To make *Euler*, for example, the current algorithm, just type "e". The submenu will be erased, and the NUMERICS menu will reappear. Selecting from a submenu is done in the same way as selecting from one of the main menus. The only exception occurs when selecting from the submenu containing names of equations that appears after picking "Equation"; in this case, you must type in not just the first letter, but the full name of the equation of your choice, followed by <return>. As a general rule, you are expected to type the part of a menu or submenu entry that is in yellow or green.

Finally, you should keep in mind that every time PHASER starts up, the screen looks like the one in *Figure 4.2,* and that all things such as "Dimension," "Equation," "Step size," etc., are assigned their default values, as explained in section 4.1. Almost all of these values remain the current values until changed explicitly. One important exception should be noted: if you change the current "Dimension," then a new equation will become the current one, replacing "linear2d".

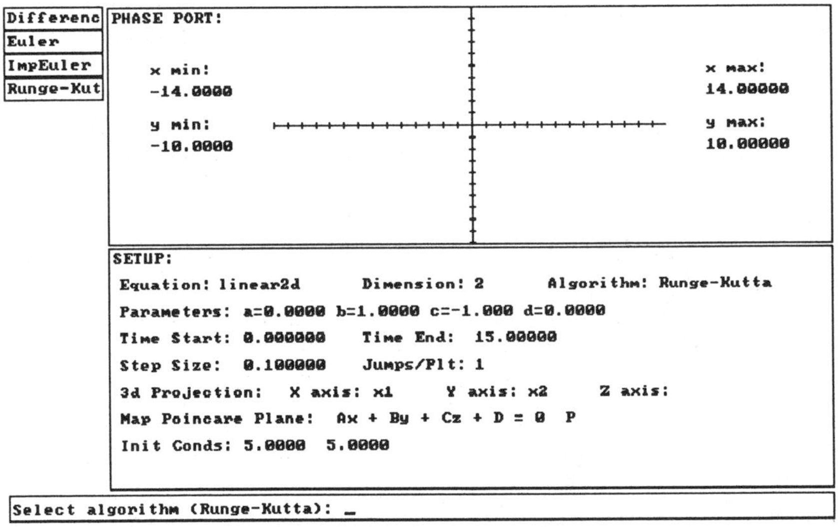

Differenc
Euler
ImpEuler
Runge-Kut

PHASE PORT:
x min: -14.0000
y min: -10.0000
x max: 14.00000
y max: 10.00000

SETUP:
Equation: linear2d Dimension: 2 Algorithm: Runge-Kutta
Parameters: a=0.0000 b=1.0000 c=-1.000 d=0.0000
Time Start: 0.000000 Time End: 15.00000
Step Size: 0.100000 Jumps/Plt: 1
3d Projection: X axis: x1 Y axis: x2 Z axis:
Map Poincare Plane: Ax + By + Cz + D = 0 P
Init Conds: 5.0000 5.0000

Select algorithm (Runge-Kutta): _

Figure 4.3. The submenu of numerical algorithms provided by PHASER.

Chapter 5
Lessons with PHASER

This chapter contains a sequence of lessons designed to teach you, step by step, how to use some of the basic capabilities of PHASER. The easiest way to learn how to use a computer program is to experience it first-hand. Therefore, you should work through this chapter with a finger on the keyboard.

The commands and inputs you are expected to type are listed in **boldface** in the left column. The corresponding responses of PHASER and explanatory remarks are given on the right. After you type each command, you should wait long enough for it to be executed. Various helpful messages may come up during simulations. Make sure you read them.

If you have not done so already, please read the instructions in **Read-Me Chapter** on how to install and run PHASER on your IBM Personal Computer. In case you are using the CGA version of PHASER, before starting PHASER, you should first execute the DOS graphics function by typing

graphics <ret>

Here <ret> is the return or the <enter> key, the large one with an arrow, on the right. This function allows you to print the contents of the screen on the graphics printer at any time during simulations. This is done by pressing the special <Print Screen>, while holding the <Shift> key. All the illustrations in this book are made using this printing facility. If you are using the EGA/VGA version of PHASER, please consult **Read-Me Chapter** on how to print the screen on either EPSON/IBM dot matrix or HP LajerJet+ laser printer.

To run PHASER, change to the directory where it is installed, and type

phaser <ret>

After about ten seconds, a logo will appear, followed by the screen image in *Figure 4.2*. You are now ready to start the first lesson. Although each lesson is an independent unit, you should study them in sequence. If at any point you get bogged down, as a last resort hit the return key several times until one of the menus whose last entry is *"Quit"* comes up. Then type **q** followed by **y** to exit, and start afresh.

Lesson 1. *In and out of PHASER.*

In this lesson, we will learn how to enter and exit PHASER. The menu you will see on the screen is called NUMERICS, and is one of the three main menus.

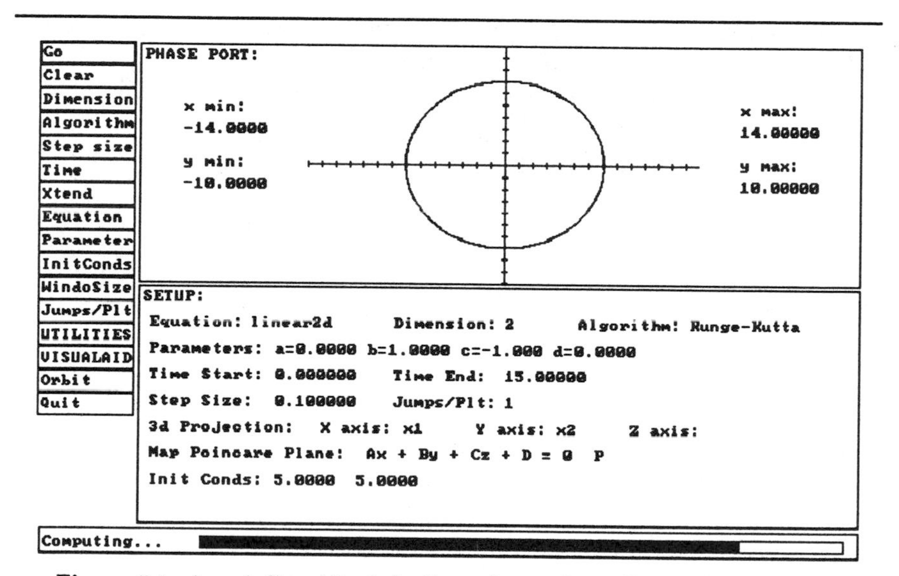

Figure 5.1. A periodic orbit of the linear harmonic oscillator. The menu on the left is the NUMERICS menu.

COMMAND: RESPONSE/EXPLANATION:

phaser <ret> To start PHASER. In about ten seconds, the screen should look like *Figure 4.2.*

g *Go*: To compute and display using the current settings (in this case, the defaults). You will see drawn the circular orbit of the linear harmonic oscillator given in equation (4.2). In the message line, a moving bar monitoring the progress of computations will appear, as shown in *Figure 5.1.*

After typing *Go*, you can halt computations temporarily by pressing <esc>, the escape key. To continue, press <return>; to abort, press <esc> again.

c *Clear*: To erase the content of the phase portrait view. The circle will disappear.

q *Quit*: To exit PHASER and go back to the operating system. You will be asked if you are sure.

y *Yes*! I do want to quit.

In the following lessons, we will assume that PHASER is in the start up configuration, and that the values of all things are set to the defaults. If you are not certain that you are in this position, exit and reenter PHASER before each lesson, as described above.

Lesson 2. *Changing menus.*

In this lesson, we will meet the second main menu, called VISUALAID, and toggle several graphical aids.

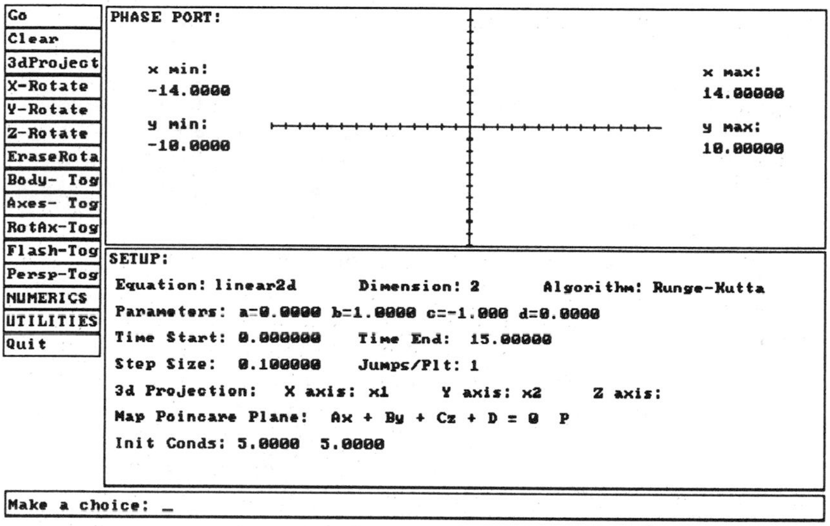

```
Go          PHASE PORT:
Clear
3dProject     x min:                                    x max:
X-Rotate      -14.0000                                  14.00000
Y-Rotate      y min:                                    y max:
Z-Rotate      -10.0000                                  10.00000
EraseRota
Body- Tog
Axes- Tog
RotAx-Tog
Flash-Tog   SETUP:
Persp-Tog   Equation: linear2d     Dimension: 2     Algorithm: Runge-Kutta
NUMERICS    Parameters: a=0.0000 b=1.0000 c=-1.000 d=0.0000
UTILITIES   Time Start: 0.000000   Time End: 15.00000
Quit        Step Size: 0.100000    Jumps/Plt: 1
            3d Projection:  X axis: x1    Y axis: x2    Z axis:
            Map Poincare Plane:  Ax + By + Cz + D = 0  P
            Init Conds: 5.0000  5.0000

Make a choice: _
```

Figure 5.2. The VISUALAID menu.

COMMAND: RESPONSE/EXPLANATION:

g	*Go*: See Lesson 1.
c	*Clear*: See Lesson 1.
v	*VISUALAID*: To bring up the VISUALAID menu. The current menu will be erased, and the VISUALAID menu will immediately be drawn in the same place; see *Figure 5.2.*
a	*Axes-Tog*: To toggle the axes on/off. By default they are on, as indicated by the highlighting of the word *Tog* in red. Therefore, the axes will now be erased from the Phase portrait view, and the highlighting will be turned off.
a	*Axes-Tog*: The axes will be put back in the phase portrait view, and *Tog* will be highlighted in red.

f	*Flash-Tog*: To toggle a flashing marker off/on. Since by default it is off, it now will be turned on (notice the highlighting of *Tog* on the menu).
g	*Go*: The same circular orbit will be drawn with a flashing marker at the current point being plotted.
n	*NUMERICS*: To bring up the NUMERICS menu.
x	*Xtend*: To extend the orbit. You will be asked for a new *End* time.
30 <ret>	The new *End* time is 30 (the *Set up* view is updated to show this). The orbit will be computed from its last point until time 30. Notice that that you do not have to press *Go*. The bar will be rescaled according to the new *End* time.
c	*Clear*: The circle will be removed.
t	*Time*: To change the current start and end times.
<ret>	To keep the start time the same (0.0).
15 <ret>	To make the end time 15.0. Notice that the *Set up* view is updated to display the new times. At this point you can either continue to the next lesson, or exit PHASER by executing the following two commands:
q	*Quit.*
y	*Yes*!

Lesson 3. *Direction field.*

In this lesson, we will see the third main menu, the UTILITIES menu, and draw the direction field of a linear harmonic oscillator.

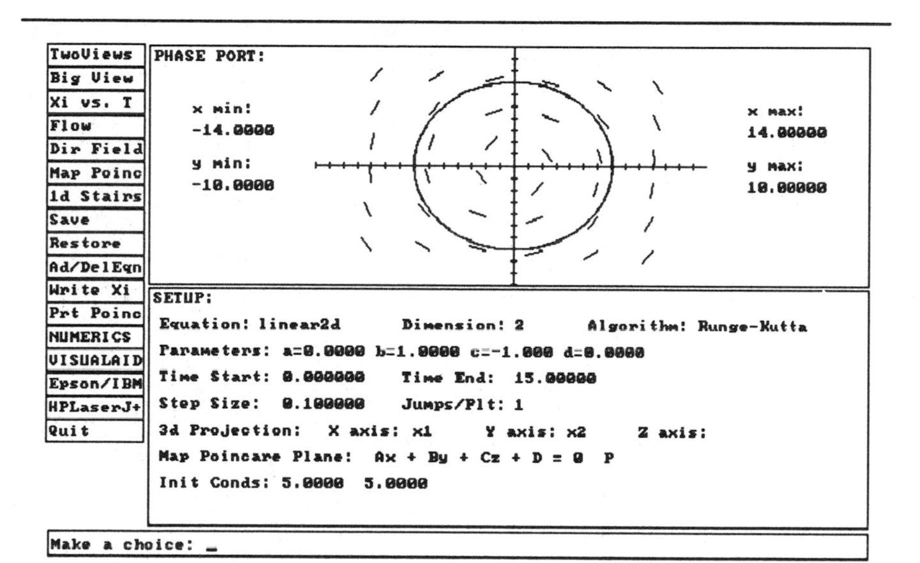

Figure 5.3. Direction field and an orbit of the linear harmonic oscillator given by equation (4.2). The menu on the left is the UTILITIES menu.

COMMAND:	RESPONSE/EXPLANATION:
u	*UTILITIES*: To bring up the UTILITIES menu.
d	*DirField*: To draw the direction field (in dimension two *only*) of the current equation, (4.2). You will be asked to specify a grid size.
6 <ret>	The grid size is 6. A 6×6 grid of equally spaced green dots will appear. Then, starting from each dot, a (normalized) vector along the vector field of equation (4.2) will be drawn.
g	*Go*: To draw the same circular orbit. Notice that although the UTILITIES menu does not contain *Go* or *Clear*, you can still use them (this is not true for any other entry). At this point the screen should look like *Figure 5.3.* If you have continued from the previous

	lesson, the flasher should still be on.
c	*Clear*: The contents of the Phase portrait view will be erased.
n	*NUMERICS*: To bring back the NUMERICS menu.

At this point, you can either continue to the next lesson or exit PHASER, as described in Lesson 1.

Lesson 4. *Changing views.*

In this lesson, we will first learn how to enlarge the picture from Lesson 3, and then how to return to the original screen configuration.

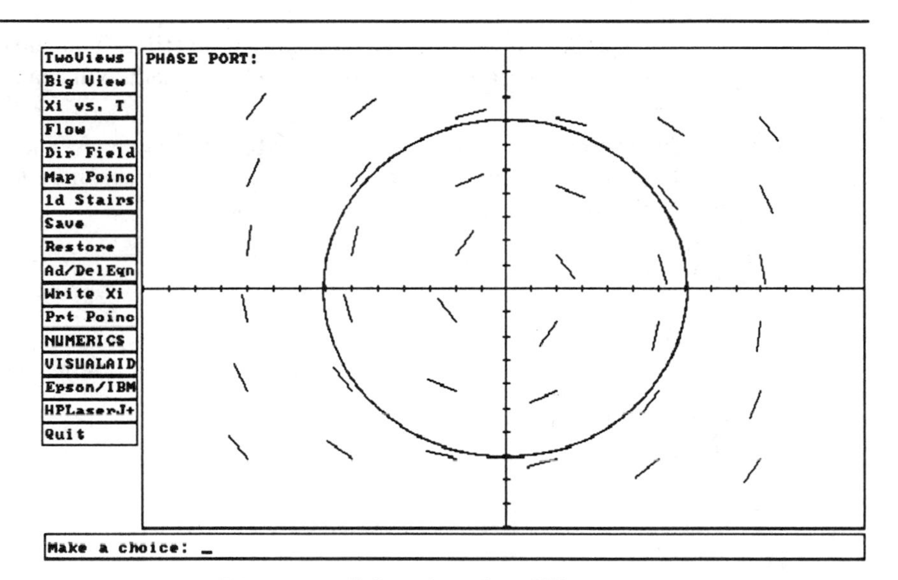

Figure 5.4. Enlarged version of Figure 5.3.

COMMAND:	RESPONSE/EXPLANATION:
u	*UTILITIES*: To bring up the UTILITIES menu.
b	*Big View*: To select a view for the entire viewing area. A submenu containing the names of nine possible views will appear, and you will be prompted to select one of them to enlarge.
p	*PhasePort*: To blow up the Phase portrait view. The submenu of views will be erased, and the previous main menu, the UTILITIES menu, will be redisplayed.
d	*DirField*: To draw the direction field; see Lesson 3.
6 <ret>	The grid size is 6; see Lesson 3.
g	*Go*: To draw the orbit; see Lesson 3. Now the screen should look like *Figure 5.4*.

t	*TwoViews*: To display two new views. The submenu containing the names of nine possible views will come up and you will be asked to select the top view.
p	*PhasePort*: To put phase portrait on the top half of the viewing area. You will be asked to select the bottom view.
s	*Set up*: To put the *Set up* view on the bottom. The submenu will be erased, and the UTILITIES menu will reappear.
n	*NUMERICS*: To bring up the NUMERICS menu. We are back at the beginning again. You can safely continue to the next lesson without exiting PHASER.

Lesson 5. *Graphing variables.*

In this lesson, we will learn how to draw simultaneously the graphs of the dependent variables x_i versus time, and the orbit of equation (4.2) on the phase plane.

COMMAND:	RESPONSE/EXPLANATION:
u	*UTILITIES*: To bring up the UTILITIES menu.
t	*TwoViews*: To display two new views.
x	*Xi vs. T*: To put the graph of the ith variable versus time as the top view. By default i is 1; that is, x1 vs. time will be plotted. The minimum and the maximum values of x1 are shown on the right. The time axis runs from the *Start* time to the *End* time (0 to 15.0, in this case). We will see shortly how to plot x2 vs. time.
p	*PhasePort*: To set the phase portrait as the bottom view. The UTILITIES menu will reappear.
g	*Go*: To plot x1 vs. time in the top view, and x1 vs. x2 in the bottom view. The screen should now look like *Figure 5.5.*
x	*Xi vs. T*: To specify the i of a different dependent variable xi, and the minimum and the maximum values of the variable. Notice that *Xi vs. T* appears both on the submenu of views and on the UTILITIES menu, but that the two entries do different things.
2 <ret>	To plot x2 vs. time.
<ret>	To keep the minimum of the *Xi vs. T* view the same (-14.0).
<ret>	To keep the maximum of the *Xi vs. T* view the same (14.0). Notice that the label in the *Xi vs. T* view has been changed from x1 to x2.
g	*Go*: To plot x2 vs. time in the top view, and x1 vs. x2 in the bottom view. At this point the screen should look like *Figure 5.6.*

You should finish this lesson by repeating the last four steps of Lesson 4 to return to the original screen configuration, as in *Figure 4.2.* Otherwise, exit and reenter PHASER before the next lesson.

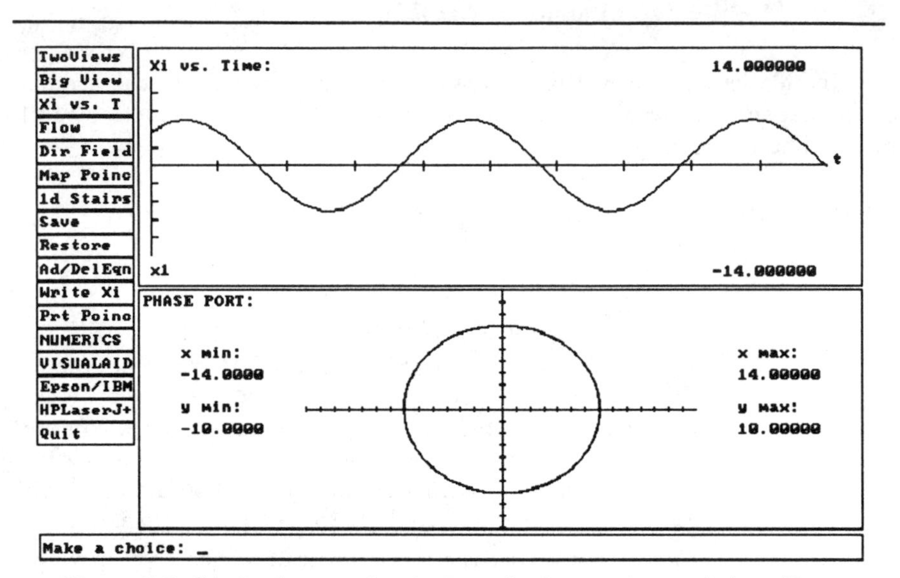

Figure 5.5. Graph of x_1 vs. time is shown in the top view, and the orbit x_1 vs. x_2 in the bottom view.

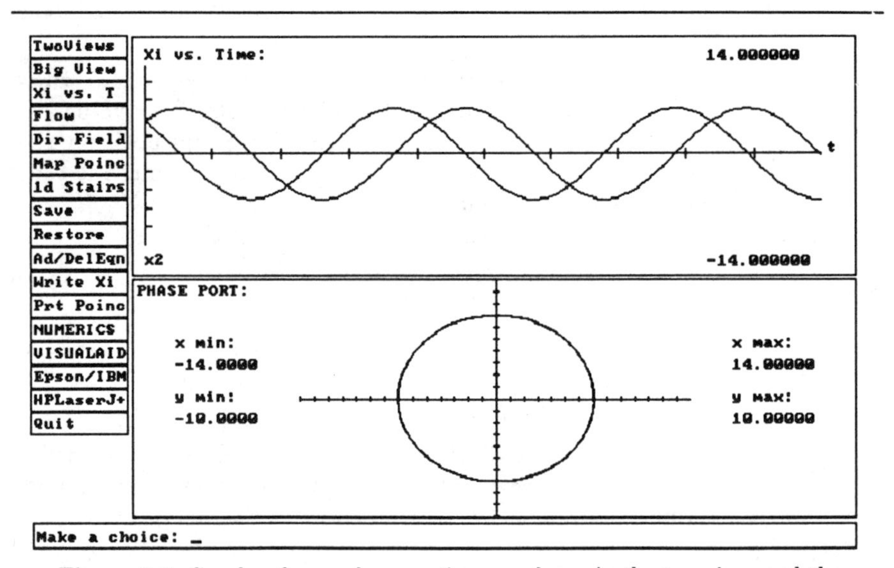

Figure 5.6. Graphs of x_1 and x_2 vs. time are shown in the top view, and the orbit x_1 vs. x_2 in the bottom view.

Lesson 6. *Initial conditions and parameters.*

In this lesson, we will first learn how to enter multiple sets of initial conditions whose corresponding solutions will be plotted simultaneously. Then we will change some of the parameters in equation (4.1), *linear2d*.

COMMAND:	RESPONSE/EXPLANATION:
i	*InitConds*: To enter new initial conditions. You will be asked to enter the value of x1. Notice that the current value, which is 5.0, is shown in parentheses.
<ret>	To keep the current value of x1 (5.0). You will then be asked to enter x2.
<ret>	To keep the current value of x2 (5.0). You will be asked if you want to enter another set of initial conditions.
y	*Yes*, I do. You will be asked to specify x1 again.
3 <ret>	The value of x1 is 3.0. You will be asked to enter x2 again.
3 <ret>	The value of x2 is 3.0. You will be asked if you wish to enter another set of initial conditions.
n	*No*: To stop entering any more initial conditions. Notice that the last line of the *Set up* view is updated to display the new entries. From now on, these two sets of initial conditions will be used in computations until changed. (You can enter up to six sets of initial conditions whose corresponding solutions are plotted simultaneously, but the last four will not be displayed in the *Set up* view unless it is enlarged.)
g	*Go*: You should now see the first solution in blue and the second one in red (*Figure 5.7*). This is the phase portrait of a *linear center*. Observe that the solutions are in phase.
p	*Parameter*: To change parameters in the definition of the current equation. A submenu containing the four parameters in the equation *linear2d* (4.1) will appear.
a	To change the value of the parameter *a* in *linear2d*. Note that the current value (0.0) is shown in parentheses.
-0.2 <ret>	The value of *a* is now -0.2. The second line of the *Set up* view will be updated to show the new value.
d	To change the value of the parameter *d* in *linear2d*.

-0.2 <ret> The value of d is now -0.2. The second line of the *Set up* view will be updated to show the new value.

<ret> To stop changing parameters and return to the NUMERICS menu.

c *Clear*: To remove the contents of the *PhasePort* view.

g *Go*: Now, you should see the two solutions of the system,

$$x_1' = -0.2x_1 + x_2$$

$$x_2' = -x_1 - 0.2x_2 ,$$

starting from (5, 5) and (3, 3), and spiraling towards the origin, as illustrated in *Figure 5.8*.

In preparation for the next lesson, you should either change all things to their default values, or exit PHASER and reenter.

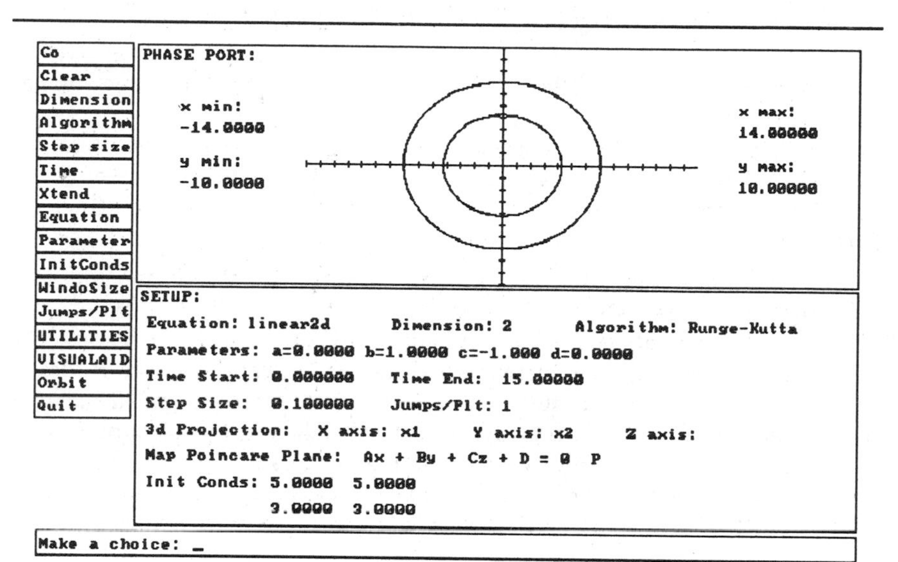

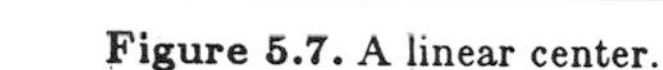
Figure 5.7. A linear center.

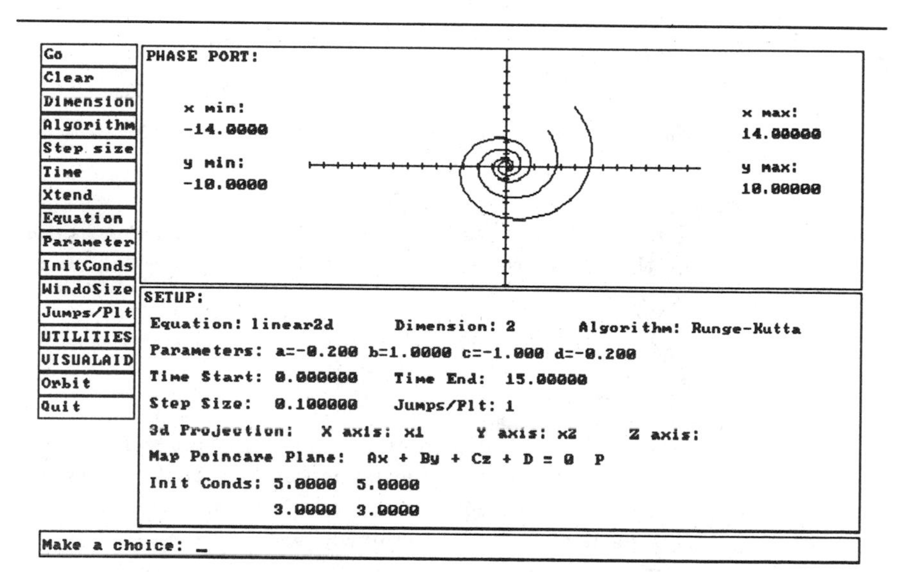

Figure 5.8. A linear spiral sink.

Lesson 7. *Making transcripts for demonstrations.*

In this lesson, we will learn how to save a sequence of screen images in a transcript, and how to play it back later. This is an extremely powerful facility enabling instructors to prepare "electronic lectures" for classroom demonstrations. It is also useful for saving the current environment in a form that can be reloaded into PHASER at a later time for further work.

At all times, PHASER keeps track of every bit of information that is necessary to recreate the current picture on the screen. When you *Save*, all the current settings of the step size, the initial conditions, the equation, etc., and all the current views in the viewing area are recorded in a user-specified file on the disk. When you *Restore* from an existing file, all this information is reloaded into PHASER and the picture is computed, as if the menu entry *Go* were activated. See Lesson 15, the *Grand Finale*, and section 6.2 for more information on these rather unique capabilities of PHASER. *Note*: Direction fields and flows cannot be saved.

COMMAND:	RESPONSE/EXPLANATION:
u	*UTILITIES*: To bring up the UTILITIES menu.
g	*Go*: You should see the familiar periodic orbit traced.
s	*Save*: To save the current settings and the window configuration. You will be asked for a file name.
lesson7 <ret>	The file name is *lesson7* (no spacing). All necessary information to recreate this picture will be recorded in file *lesson7*.
b	*Big View*: To enlarge a view.
p	*PhasePort*: The phase portrait will be the current big view.
s	*Save*: To save the current view and the settings. You will again be asked for a file name. (Notice that you do not have to press *Go* before *Save* unless, of course, you would like to see what is being saved.)
lesson7 <ret>	Since a file with this name already exists, you will be asked whether to *append* or to *overwrite* (a/o).
a	To append the current view and the settings to file *lesson7*. Now, there are two pictures in file *lesson7*.
r	*Restore*: To play back an existing transcript file, created using *Save*. You will be asked for a file name.

lesson7 <ret>

To play back what is in the file *lesson7*. You will first see the familiar picture shown in *Figure 5.1.* For the second picture, press the space bar to see the enlarged version of the phase portrait. At this point, all settings such as step size, equation, views, etc., are the ones belonging to the last picture.

q *Quit.*

y *Yes.*

If you wish, you can now look at the content of file *lesson7* by issuing the DOS command **type lesson7 <ret>**. Do not try to decipher it, however. The file can be erased from the disk using the DOS command **erase lesson7 <ret>**.

Lesson 8. *Browsing through the library.*

In this lesson, we will learn how to select an equation from the library of PHASER and display its text. To access a specific equation, you must know its dimension and whether it is a differential or a difference equation. Naturally, the dimension can be specified by the *Dimension* entry on the NUMERICS menu. The differential/difference type is a bit more subtle: first, you must select the *Algorithm* entry of the NUMERICS menu. Then, from the submenu that comes up, pick *Difference* if you want a difference equation; otherwise, select any one of the remaining three algorithms (*Euler, ImpEuler*, or *Runge-Kut*) for a differential equation. Once the desired dimension and the differential/difference type are specified, a default equation is loaded. Now, you can specify the equation of your choice by first picking the *Equation* entry of the NUMERICS menu, and then selecting your equation from the submenu that comes up.

```
TwoViews     EQUATION:
Big View                  linear2d
Xi vs. T          Linear system in dimension two
Flow
Dir Field              x1' = a * x1 + b * x2
Map Poinc              x2' = c * x1 + d * x2
1d Stairs
Save
Restore
Ad/DelEqn
Write Xi     SETUP:
Prt Poinc     Equation: linear2d      Dimension: 2      Algorithm: Runge-Kutta
NUMERICS      Parameters: a=0.0000 b=1.0000 c=-1.000 d=0.0000
VISUALAID     Time Start: 0.000000   Time End:  15.00000
Epson/IBM     Step Size:  0.100000   Jumps/Plt: 1
HPLaserJ+     3d Projection:   X axis: x1     Y axis: x2     Z axis:
Quit          Map Poincare Plane:  Ax + By + Cz + D = 0  P
              Init Conds: 5.0000  5.0000

Make a choice: _
```

Figure 5.9. The text of *linear2d* and the *Set up* view.

COMMAND:	RESPONSE/EXPLANATION:
u	*UTILITIES*: To bring up the UTILITIES menu.

t *TwoViews*: To display two new views.

e *Equation*: To put the text of the current equation in the top view. You should now see the text of *linear2d* displayed (*Figure 5.9*). The prime stands for derivative in the case of ordinary differential equations, and the next iterate in the case of difference equations.

s To keep the bottom view the same (*Set up*). Current values of the parameters a, b, c and d in *linear2d* are visible on the second line of the *Set up* view.

n *NUMERICS*: To bring up the NUMERICS menu.

e *Equation*: To select a new equation. A long submenu containing the names of all the two-dimensional differential equations in the library will come up.

pendulum <ret>

To make *pendulum* the current equation. Notice that the full name of the equation, exactly as it appears on the submenu, must be typed, followed by <return>. The submenu of equations will be erased, and the NUMERICS menu will be redisplayed. Observe that the *Set up* view is updated to display the parameters and their default values in the new equation. All the other settings remain the same. By repeating the previous three steps, you can look at all the two-dimensional differential equations in the PHASER library. However, if there are more than nineteen equations in the current dimension and differential/difference type, then the last entry on the submenu will be *MORE*. To go to the next page of equations, you should type *MORE <ret>*.

a *Algorithm*: To change the algorithm. A short submenu of available algorithms for differential and difference equations (maps) will appear.

d *Difference*: To make *difference* the current algorithm. Now, as above, you can look at all the two-dimensional difference equations (maps) in the library. Notice that the default difference equation for dimension two, *dislin2d*, the two-dimensional discrete linear system, is loaded (there is a different default equation in each dimension and differential/difference type). In addition, the step size is set to 1.

e *Equation*: To select a new equation. A long submenu containing the names of all the two-dimensional difference equations will come up.

henon <ret>	To make *henon* the current equation. The text of this equation and the parameter settings will be displayed in the appropriate views.
a	*Algorithm*: To change the algorithm.
r	*Runge-Kut*: To change the current algorithm back to Runge-Kutta. Notice that the two-dimensional default equation of differential type, *linear2d*, is made the current equation. Moreover, the step size is set to the default value 0.1.
d	*Dimension*: To change the dimension.
3 <ret>	Now the current dimension is 3. Observe that the current equation is changed to *lorenz*, which is the default equation in three dimensions when the current algorithm is of differential type (see section 7.3 for more information on the Lorenz equation). In general, when the dimension is changed, the default equation in the new dimension and the algorithm type are loaded automatically. At this point, if you wish, you can look at all the three-dimensional differential equations.
d	*Dimension*: To change the dimension.
2 <ret>	The current dimension is now 2, and the algorithm is still of differential type. Therefore, we are back to the differential equation *linear2d*.

Using Lesson 4, you should change the views back to the default configuration, as in *Figure 4.2*.

Lesson 9. *Entering your equations.*

In this lesson, we will learn how to add/delete an equation to/from the library of PHASER. Once entered successfully, a new equation will be indistinguishable, as far as the user is concerned, from the permanent equations. It can be accessed by following the instructions given in Lesson 8. Since entering a new equation is at first a somewhat complicated procedure, you may want to consult concurrently the *Ad/DelEqn* entry in section 6.2, where more detailed information on this powerful facility of PHASER can be found.

We will use the following example to illustrate the process of entering a new equation into PHASER:

$$y' = 1.0 + a\,(y - t\,)^2 \qquad y\,(0) = 0.5\,, \tag{5.1}$$

where a is a parameter. Initially, we will take the value of the parameter a to be 1.0. If desired, this value can be changed interactively later, during simulations. First, you must convert the equation above into an autonomous first-order system. Using the results of Chapter 1, it is easy to see that the desired equivalent system is the following:

$$x_1' = 1.0 + a\,(x_1 - x_2)^2 \tag{5.2}$$

$$x_2' = 1.0,$$

with the initial conditions

$$x_1(0.0) = 0.5, \qquad x_2(0.0) = 0.0\;.$$

Before entering new equations, a small amount of preparation is required. You must first decide on the number, the names, and the default values of parameters in your equations. In the definition of equations, *parameters* are single letters whose values can be changed interactively during simulations. For example, in *linear2d* (4.1) there are four parameters: *a, b, c,* and *d.* In equation (5.2), a is the only parameter. Second, you must determine the number and the values of the constants in your system. *Constants* are real numbers appearing as factors, exponents, etc. on the right-hand side of your equations. The constants will be named c1, c2, ..., c9. Giving them names is for the convenience of entering your equations since once entered, all constants will be replaced by their values. Their names can then be safely forgotten. In equation (5.2), for instance, there are two constants; we will call them c1 = 1.0 and c2 = 2.0.

When entering the right-hand side of your equations, the allowable symbols are x1, ..., x9 for variables; single letters for parameters; c1, ..., c9 for constants; + for addition; - for subtraction; * for multiplication; / for division; ^ for power; left and right parentheses. Several common functions such as abs(), cos(), sin(), exp() are also provided.

COMMAND:	RESPONSE/EXPLANATION:
u	*UTILITIES*: To bring up the UTILITIES menu.
a	*Ad/DelEqn*: To add/delete your own equation to the library. (Entries in the permanent library cannot be deleted.) You will be asked whether you want to add or delete (a/d).
a <ret>	To add an equation. You will be asked to enter the name of your equation. A maximum of eight characters is allowed.
mine <ret>	The name of the equation (5.2) will be *mine*. PHASER requires the names of the equations in the library to be unique. Therefore, if someone else has tried this lesson before, you will get a message saying "*mine* has already been defined; delete old one before adding new eqn". You will then be given another chance to provide a new name. At this point, the easiest thing to do is to call your equation by another name, say *mine1*. If you are curious how to delete an equation, see section 6.2. Next, you will be asked for the type of *mine*: 0 = difference, 1 = differential.
1 <ret>	The system (5.2) is a differential equation. You will be asked to enter its dimension.
2 <ret>	The dimension of *mine* is 2. You will be asked the number of parameters in *mine*.
1 <ret>	There is one parameter in *mine*. You will be asked the name of the parameter in *mine*.
a <ret>	The name of the parameter is *a*. You will be asked the default value of the parameter *a*.
1.0 <ret>	The default value of *a* is 1.0. You will be asked the number of constants in *mine*.
2 <ret>	There are two constants. You will be asked to enter the value of the first constant c1.
1.0 <ret>	The value of c1 is 1.0. You will be asked to enter the value of the second constant c2.
2.0 <ret>	The value of c2 is 2.0. You will be asked to enter the formula for the derivative of the first variable: $x1' =$.
c1 + a*(x1 - x2)^c2 <ret>	Notice that this is precisely the right-hand side of the first equation in the system (5.2). You will be asked to enter the formula for the derivative of the second

variable: $x2' =$.

cl <ret> Notice that this is precisely the right-hand side of the second equation in the system (5.2). Your equation *mine* is now added to the library, and you will be returned to the UTILITIES menu.

n *NUMERICS*: To bring up the NUMERICS menu. This step is added so that, if you wish, you can proceed to the next lesson without quitting PHASER.

This useful facility of PHASER is not without limitations, however. Be forewarned that integrating a new equation which contains long and complicated formulas tends to be a slow process.

Lesson 10. *Euler vs. Runge-Kutta.*

In this lesson, we will learn how to look at our new equation *mine*, and compare the results of the algorithms of Runge-Kutta and Euler on it (see Braun [1983], pp. 97-114). We assume that you have completed the previous lesson successfully.

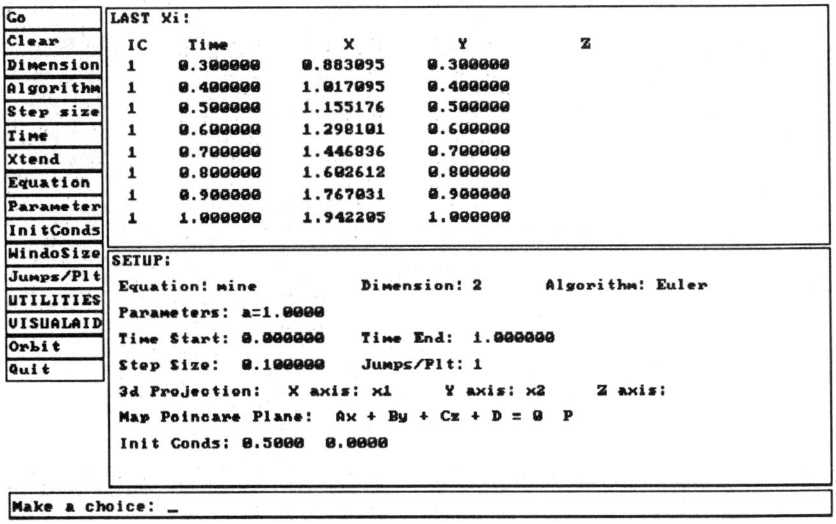

```
Go          | LAST Xi:
Clear       |   IC   Time        X           Y          Z
Dimension   |   1    0.300000    0.883095    0.300000
Algorithm   |   1    0.400000    1.017095    0.400000
Step size   |   1    0.500000    1.155176    0.500000
Time        |   1    0.600000    1.298101    0.600000
Xtend       |   1    0.700000    1.446836    0.700000
Equation    |   1    0.800000    1.602612    0.800000
Parameter   |   1    0.900000    1.767031    0.900000
InitConds   |   1    1.000000    1.942205    1.000000
WindoSize   | SETUP:
Jumps/Plt   |  Equation: mine          Dimension: 2       Algorithm: Euler
UTILITIES   |  Parameters: a=1.0000
VISUALAID   |  Time Start: 0.000000    Time End: 1.000000
Orbit       |  Step Size:  0.100000    Jumps/Plt: 1
Quit        |  3d Projection:   X axis: x1      Y axis: x2      Z axis:
            |  Map Poincare Plane:   Ax + By + Cz + D = 0   P
            |  Init Conds: 0.5000  0.0000

Make a choice: _
```

Figure 5.10. Last several computed values of the variables in equation *mine* (5.2), using *Euler* with step size 0.1.

COMMAND:	RESPONSE/EXPLANATION:
e	*Equation*: To bring up the submenu of 2D differential equations.
MORE <ret>	To go to the next page of the submenu of names of equations.
mine <ret>	To make *mine* the current equation.
u	*UTILITIES*: To bring up the UTILITIES menu.
t	*TwoViews*: To display two new views.
e	*Equation*: To put the text of the current equation as the top view. The text of *mine* should be displayed. Make

Key	Action
	certain that the text is correct; otherwise, repeat the previous lesson.
<ret>	To keep the the bottom view the same (*Set up*).
t	*TwoViews*: To display two new views.
l	*Last Xi*: (Letter "el," not one) To put this view, which will display the last eight iterations of time and the dependent variables xi, on top.
<ret>	To keep the bottom view the same (*Set up*).
n	*NUMERICS*: To bring up the NUMERICS menu.
t	*Time*: To change the current time interval. You will be asked for the *Start* time.
<ret>	To keep the Start time the same (0.0). You will be asked for the *End* time.
1.0 <ret>	The End time is 1.0.
i	*InitConds*: To enter new initial conditions.
0.5 <ret>	The value of x1 is 0.5.
0 <ret>	The value of x2 is 0.0.
<ret>	To stop entering initial conditions (of course, **n** does this also).
g	*Go*: You should now see the last several computed values of time, x1 (under x) and x2 (under y) using *Runge-Kutta* with step size 0.1 (cf. Braun [1983], p. 113). Notice that the values of Time and x2 are the same because the system (5.2) comes from a nonautonomous equation (5.1).
a	*Algorithm*: To change the algorithm. A submenu containing the names of available algorithms will appear.
e	*Euler*: To make *Euler* the current algorithm.
g	You should now see (*Figure 5.10*) the computed values of the variables using *Euler* with the same step size 0.1 (cf. Braun [1983], p. 99).
t	*Time*: To change the current time interval in preparation for the next lesson.
<ret>	To keep the start time the same (0.0).
15 <ret>	To make the end time 15.0.

The exact solution of the initial-value problem (5.1) is

$$y(t) = t + \frac{1.0}{2.0 - t}.$$

Compare the exact value $y(1.0) = 2.0$ with the numerical results. You can dispense with studying the phase portrait of this system; it is not very exciting.

Lesson 11. *Stair step diagrams.*

In this lesson, we will simulate the one-dimensional difference equation, *logistic*, using stair step diagrams. As our example we will use the logistic equation (3.6) given by the formula

$$x_1{}' = ax_1(1.0 - x_1)\,, \tag{5.3}$$

where prime denotes the next iterate. If you do not recall what a stair step diagram is, then you should read section 3.1 before starting the lesson.

COMMAND:	RESPONSE/EXPLANATION:
u	*UTILITIES*: To bring up the UTILITIES menu.
t	*TwoViews*: To display two views.
1	*1d Stairs*: To make the one-dimensional stair step diagram the top view.
<ret>	To keep the bottom view the same (*Set Up*).
n	*NUMERICS*: To bring up the NUMERICS menu.
d	*Dimension*: To change the dimension.
1 <ret>	To make the current dimension 1.
a	*Algorithm*: To change the algorithm.
d	*Difference*: To make the current algorithm *difference* (for iterating difference equations). Notice that the current equation is made *logistic*, which is the default equation for one-dimensional difference equations.
w	*WindoSize*: To change the window size.
-0.1 <ret>	To make x-min -0.1.
1.1 <ret>	To make x-max 1.1.
-0.1 <ret>	To make y-min -0.1.
1.1 <ret>	To make y-max 1.1.
i	*InitConds*: To change initial conditions.
0.3 <ret>	To make $x_1^0 = 0.3$.
<ret>	To stop entering any more initial conditions (of course, **n** does this also).
g	*Go*: You should see *Figure 5.11* on the screen.
c	*Clear*: To erase the picture.
u	*UTILITIES*: To bring up the UTILITIES menu.

1 *1dStairs*: To specify the power (iterate) of the current one-dimensional difference equation (***logistic***). Notice that *1d Stairs* appears both in the submenu of the view names and in the UTILITIES menu, but that the two entries do different things.

2 <ret> The second power (iterate) of the logistic map will be the current equation.

g *Go*: You should see *Figure 5.12* on the screen.

1 *1dStairs*: To specify the power of the current equation.

1 <ret> The power is 1 again. Therefore, the logistic map itself is the current equation.

b *Big View*: To enlarge a view.

1 *1dStairs*: Big view is the stair step diagram.

n *NUMERICS*: To bring up the NUMERICS menu.

t *Time*: To change time.

100 <ret> Start time is 100 (to cut out transients).

300 <ret> End time is 300.

g *Go*: You should see *Figure 5.13* on the screen.

c *Clear*: To clear the screen.

p *Parameter*: To change the parameter in the logistic equation.

a Parameter to be changed is a.

3.65 <ret> New value of a is 3.65.

<ret> To stop changing parameters.

g *Go*: You should see *Figure 5.14* on the screen.

q *Quit*: In preparation for the next lesson, you should definitely exit PHASER, unless, of course, you can set all things to their default values.

y *Yes*: Now you are ready for Lesson 12.

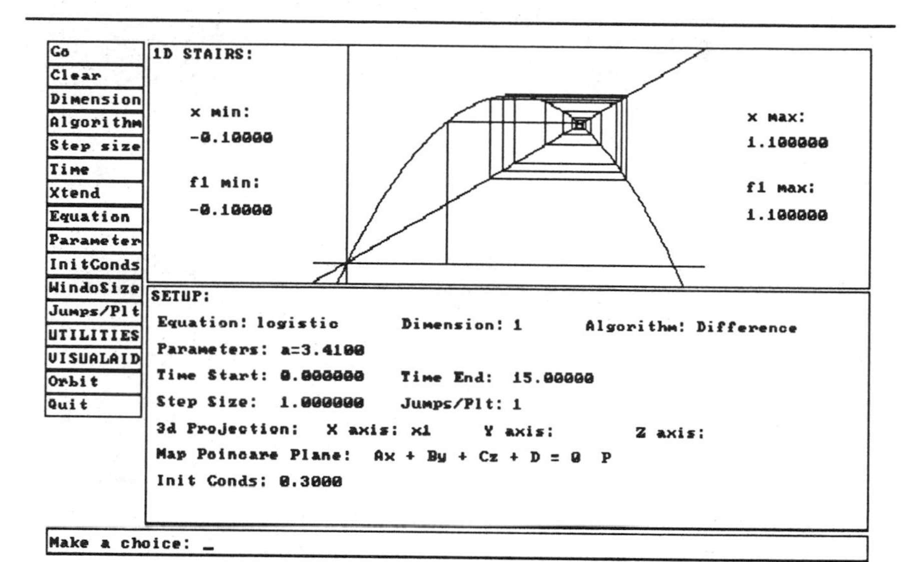

Figure 5.11. The first 15 iterates of the starting point $x^0 = 0.3$ for the logistic map, with the parameter value $a = 3.41$.

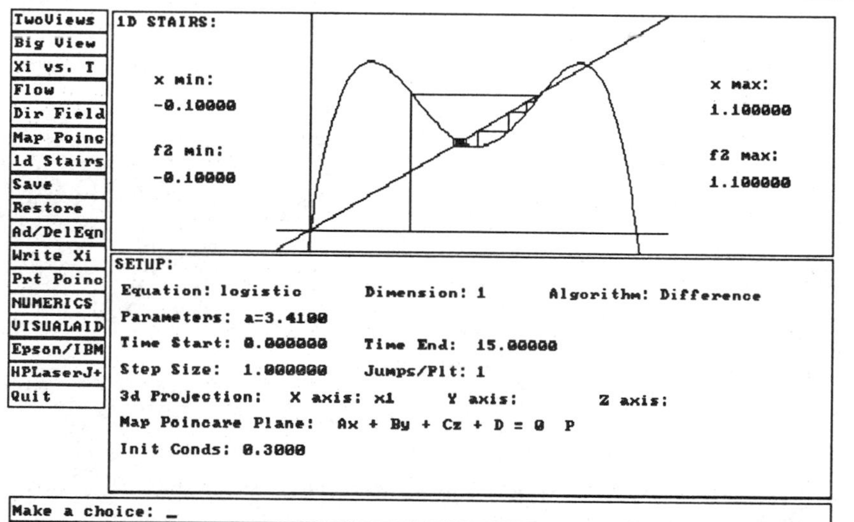

Figure 5.12. Similar to *Figure 5.11*, except that the second power, $f^{(2)}$, of the logistic map is used.

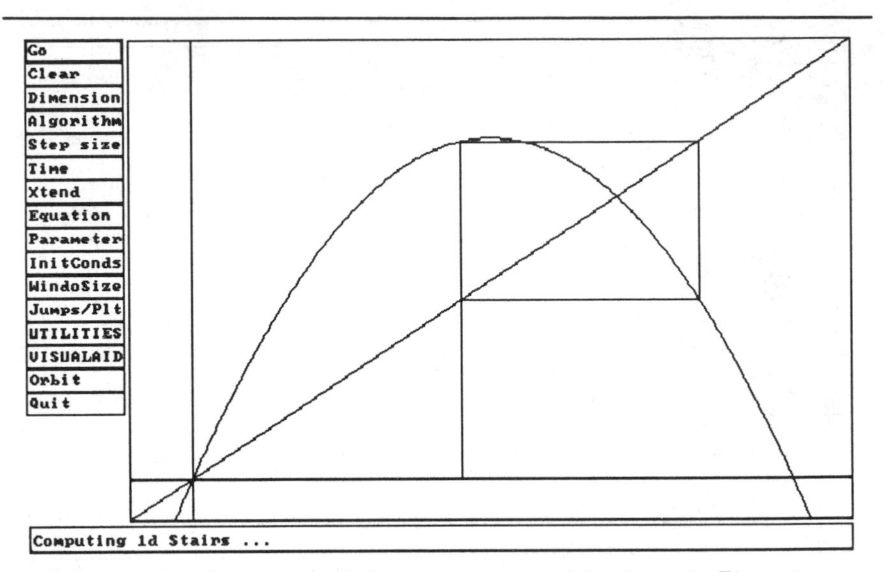

Figure 5.13. The iterates 100-300 of the same initial point as in *Figure 5.11*. Notice the attracting period-two orbit.

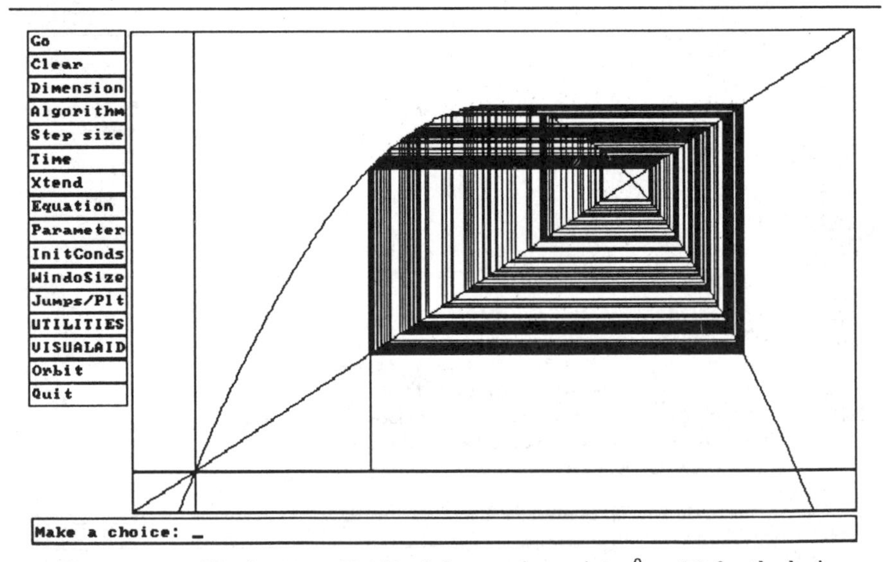

Figure 5.14. The iterates 100-300 of the starting point $x^0 = 0.3$ for the logistic map, with the parameter value $a = 3.65$.

Lesson 12. *Three-dimensional Graphics.*

In this lesson, we will get acquainted with the three-dimensional graphical facilities of PHASER. Before starting the lesson, you should read the introduction to section 6.3. Viewing a three-dimensional object on a two-dimensional screen is necessarily a complex process. Therefore, we will start with a geometrically simple object, namely a spiraling solution of the following three-dimensional linear system, *linear3d*:

$$x_1' = ax_1 + bx_2 + cx_3$$

$$x_2' = dx_1 + ex_2 + fx_3 \tag{5.4}$$

$$x_3' = gx_1 + hx_2 + ix_3 ,$$

where $a = -0.1$, $b = -1.0$, $c = 0.0$, $d = 1.0$, $e = -0.1$, $f = 0.0$, $g = 0.0$, $h = 0.0$, $i = -0.2$. Notice that for the parameter values above, the system (5.4) is decoupled: a solution spirals in the x_1, x_2 variables while it moves towards the origin in the x_3 variable.

COMMAND:	RESPONSE/EXPLANATION:
d	*Dimension*: To change the dimension.
3 <ret>	To make the current dimension 3. Notice that *lorenz*, the default differential equation in three dimensions, is made the current equation. In addition, the current (default) initial values of all the variables are 5.0.
e	*Equation*: To change the current equation.
linear3d <ret>	To make *linear3d* (5.4) the current equation.
p	*Parameter*: To change the values of the parameters in *linear3d*.
a	To change the value of the parameter a in *linear3d*.
-0.1 <ret>	To set $a = -0.1$ (recall that the old value of a was 0.0).
e	To change the value of the parameter e in *linear3d*.
-0.1 <ret>	To set $e = -0.1$ (recall that the old value of e was 0.0).
<ret>	To stop changing any more parameter values. Notice that the default values of the remaining parameters are the same as given in (5.4).

u	*UTILITIES*: To bring up the UTILITIES menu.
b	*Big View*: To select a view for the entire viewing area.
p	*PhasePort*: To make the phase portrait view the enlarged view.
v	*VISUALAID*: To bring up the VISUALAID menu.
a	*Axes-Tog*: To turn off the axes.
r	*RotAx-Tog*: To toggle on the rotated axes.
g	*Go*: Now the the orbit will be drawn in the (x_1, x_2)-plane; see *Figure 5.15.*
c	*Clear*: To clear the phase portrait view.
x	*X-Rotate*: To rotate about the (screen) x-axis. You will be asked to enter the amount of rotation in degrees.
40 <ret>	The amount of rotation about the x-axis is 40°. Observe the rotated axes in the phase portrait view.
g	*Go*: The same orbit will be recomputed and drawn in the rotated configuration; see *Figure 5.16.*
c	*Clear*: To clear the phase portrait view.
x	*X-Rotate*: To rotate about the x-axis.
50 <ret>	The amount of rotation is 50°. Since rotations are cumulative, the current rotation is 90° around the x-axis. Observe the rotated axes.
g	*Go*: The same orbit will be recomputed and drawn in the rotated configuration (in the (x_1, x_3)-plane); see *Figure 5.17.*
c	*Clear*: To clear the phase portrait view.
e	*EraseRota*: To erase all previous rotations.
g	*Go*: You should see the original spiral in *Figure 5.15.*
c	*Clear*: To clear the screen.
y	*Y-Rotate*: To rotate about the y-axis.
-80 <ret>	The amount of rotation is -80°.
x	*X-Rotate*: to rotate about the x-axis.
5 <ret>	The amount of rotation is 5°.
g	*Go*: You should now see *Figure 5.18.* Notice that the rotations are again cumulative.
c	*Clear*: To clear the screen.
e	*EraseRota*: To erase all previous rotations.
g	*Go*: You should again see the original spiral in *Figure 5.15.*

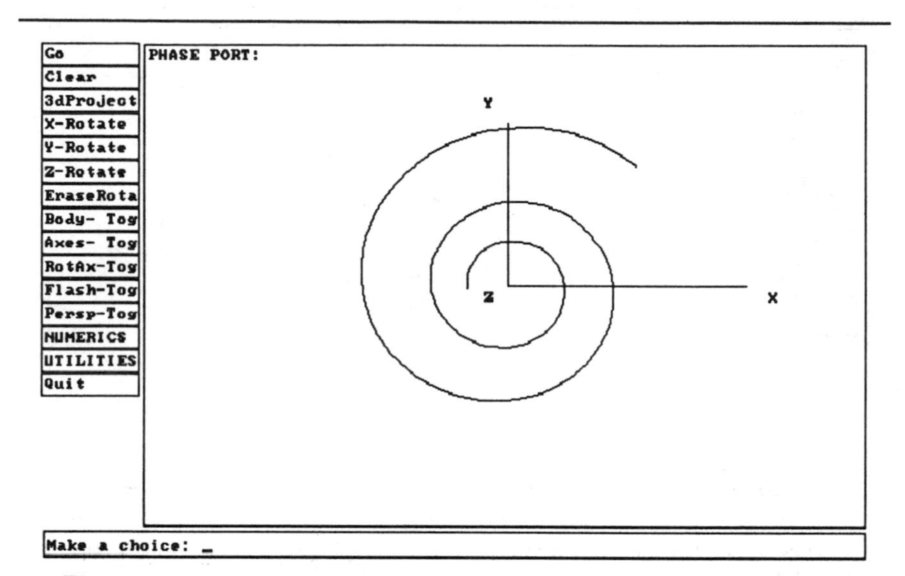

Figure 5.15. A spiraling solution of the three-dimensional linear system (5.4), as projected onto the (x_1, x_2)-plane.

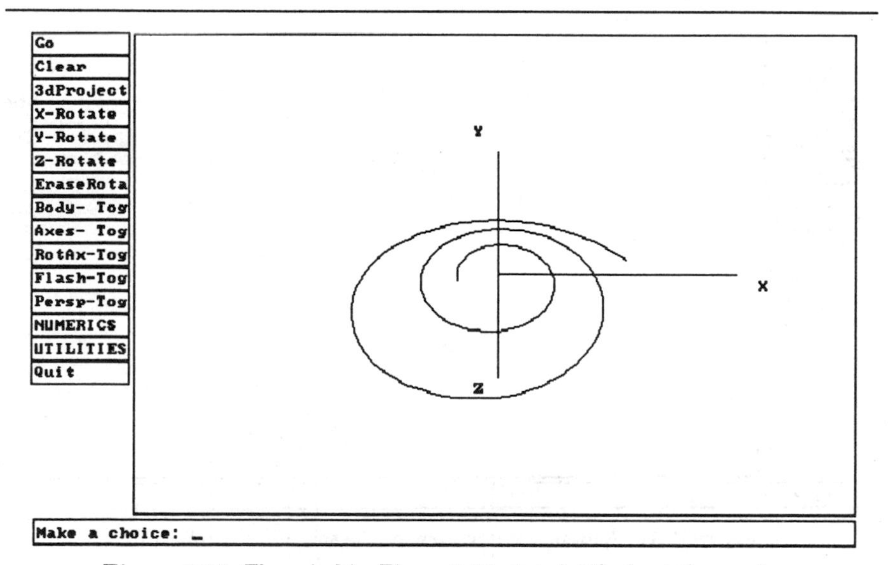

Figure 5.16. The spiral in *Figure 5.15* rotated 40° about the x-axis.

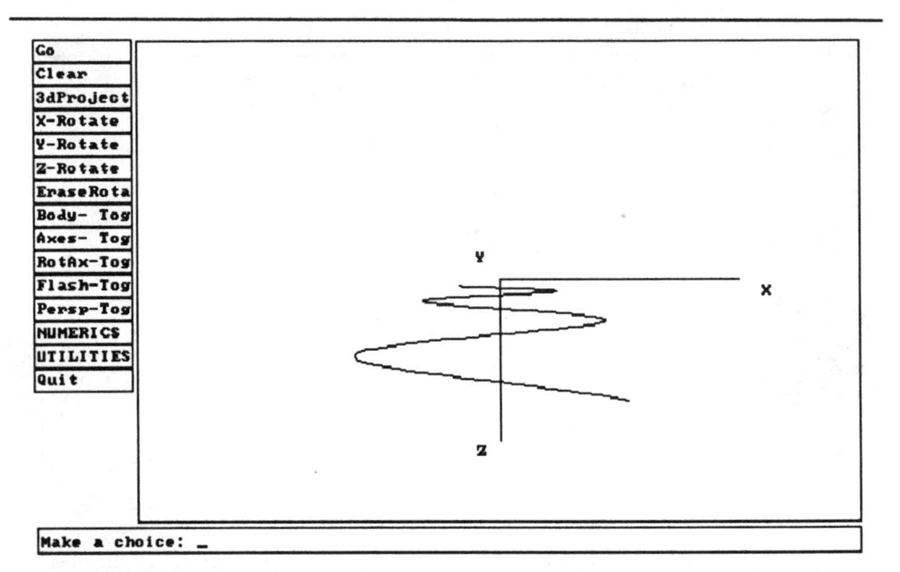

Figure 5.17. The spiral in *Figure 5.15* rotated 90°, so as to view it in the (x_1, x_3)-plane.

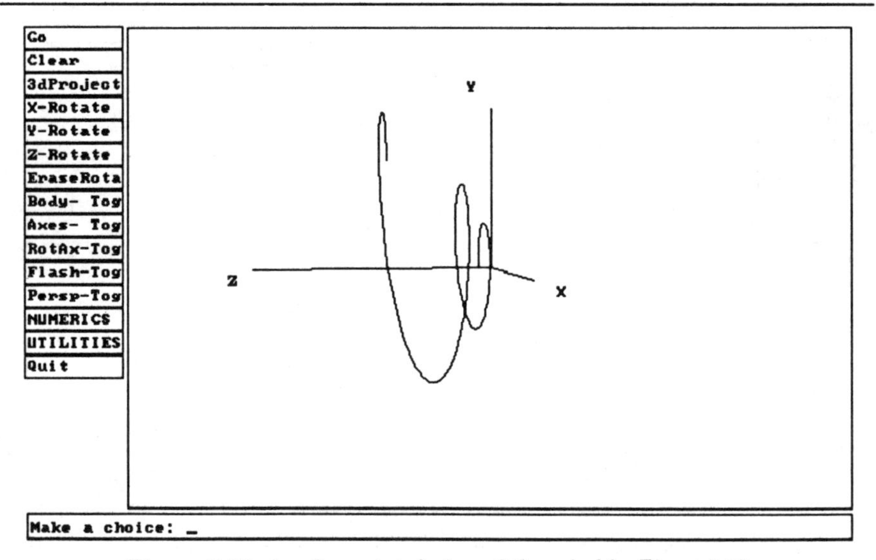

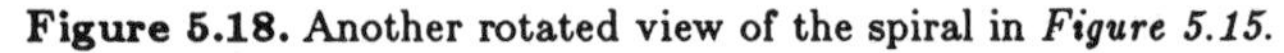

Figure 5.18. Another rotated view of the spiral in *Figure 5.15*.

Lesson 13. *Advanced graphics: A 4D example.*

In this lesson, we will explore some of the more advanced graphical facilities of PHASER. We will illustrate the concepts on a geometrically simple object -- namely, a finite cylinder in three space, one which comes from a set of four dimensional differential equations. You are encouraged to go through this lesson even if you do not completely understand the underlying mathematics.

The motion of a pair of linear harmonic oscillators can be described by the following set of differential equations:

$$\begin{aligned} x_1' &= ax_3 \\ x_2' &= bx_4 \\ x_3' &= -ax_1 \\ x_4' &= -bx_2 . \end{aligned} \tag{5.5}$$

In general, an orbit of the system (5.5) lies on a two-dimensional torus (the surface of a doughnut) in four-dimensional space. If the ratio of the frequencies a/b of the oscillators is an irrational number, then an orbit of the system (5.5) wraps around the torus many times, almost covering its entire surface.

When a two-dimensional torus in four-space is projected into three-space orthographically (by assigning the x_1, x_2, and x_3 variables to the x-, y- and z-axes, respectively, and by dropping the x_4 variable), it looks like a finite cylinder. In this lesson we will rotate such a cylinder in three dimensions, and "walk through" it using perspective viewing.

COMMAND:	RESPONSE/EXPLANATION:
d	*Dimension*: To change the dimension.
4 <ret>	To make the current dimension 4. Notice that *harmoscil*, the default differential equation in four dimensions, is made the current equation (5.5).
t	*Time*: To change the current start and end times.
<ret>	To keep the start time the same (0.0).
65 <ret>	To make the end time 65.0.
u	*UTILITIES*: To bring up the UTILITIES menu.
b	*Big View*: To select a view for the entire viewing area.
p	*PhasePort*: To make the phase portrait view the enlarged view.

v	*VISUALAID*: To bring up the VISUALAID menu.
a	*Axes-Tog*: To turn off the axes.
r	*RotAx-Tog*: To toggle on the rotated axes.
f	*Flash-Tog*: To turn on the flashing marker.
g	*Go*: Now the screen should look like *Figure 5.19.* This picture is known as a *Lissajous figure.*
c	*Clear*: To clear the phase portrait view.
x	*X-Rotate*: To rotate about the (screen) x-axis. You will be asked to enter the amount of rotation in degrees.
60 <ret>	The amount of rotation about the x-axis is 60°. Notice the rotated axes in the phase portrait view.
g	*Go*: The same orbit will be recomputed and then drawn in the rotated configuration; see *Figure 5.20.*
c	*Clear*: To clear the phase portrait view.
x	*X-Rotate*: To rotate about the x-axis.
30 <ret>	The amount of rotation is 30°. Since rotations are cumulative, the current rotation is 90° around the x-axis. Observe the rotated axes.
g	*Go*: You will get the top view of the cylinder, which is just a circle; see *Figure 5.21.* If you get tired of seeing the orbit go around the circle, hit the <esc> key twice to abort.
c	*Clear*: To clear the phase portrait view.
n	*NUMERICS*: To bring up the NUMERICS menu.
w	*WindoSize*: To change the window size. We will enlarge the boundaries of the window in preparation for *perspective* viewing. (In perspective viewing, parts of objects that are closer to the point of view look larger. Therefore, as we get very near the object, its "front" gets so large that it may fall outside the boundaries of the viewing area.)
-24 <ret>	To make x-min -24.
24 <ret>	To make x-max 24.
-20 <ret>	To make y-min -20.
20 <ret>	To make y-max 20.
<ret>	To keep z-min the same (-10).
<ret>	To keep z-max the same (10).
g	*Go*: You should see a smaller circle, as in *Figure 5.22.*

c	*Clear*: To clear the phase portrait view.
v	*VISUALAID*: To bring up the VISUALAID menu.
r	*RotAx-Tog*: To turn off the rotated axes.
p	*Persp-Tog*: To specify the z-coordinate of the viewing point, in screen coordinates, for perspective projection. (By changing the viewing point, we will "walk through" the cylinder.)
30 <ret>	The viewing point is at $z = 30$.
g	*Go*: You should now see *Figure 5.23*.
c	*Clear*: To clear the screen.
p	*Persp-Tog*: To specify the z-coordinate of the viewing point.
15 <ret>	The viewing point is at $z = 15$.
g	*Go*: You should now see *Figure 5.24*.
c	*Clear*: To clear the screen.
p	*Persp-Tog*: To specify the z-coordinate of the viewing point.
5 <ret>	The viewing point is at $z = 5$.
g	*Go*: You should now see *Figure 5.25*.
c	*Clear*: To clear the screen.
p	*Persp-Tog*: To specify the z-coordinate of the viewing point.
-5 <ret>	The viewing point is at $z = -5$.
g	*Go*: You should now see *Figure 5.26*.
p	*Persp-Tog*: To specify the z-coordinate of the viewing point.
n <ret>	*None*: No perspective. Notice that the green highlighting of *Tog* is turned off.
e	*EraseRota*: To erase all previous rotations.
c	*Clear*: To clear the screen.
g	*Go*: You should see a smaller version of the Lissajous figure in *Figure 5.19*. To recover the original picture, we will now reset the window size.
n	*NUMERICS*: To bring up the NUMERICS menu.
w	*WindoSize*: To change the window size.
-12 <ret>	To make x-min -12.
12 <ret>	To make x-max 12.

-10 <ret>	To make y-min -10.
10 <ret>	To make y-max 10.
<ret>	To keep z-min the same (-10).
<ret>	To keep z-max the same (10).
c	*Clear*: To clear the screen.
g	*Go*: You should now see the original Lissajous figure shown in *Figure 5.19*.

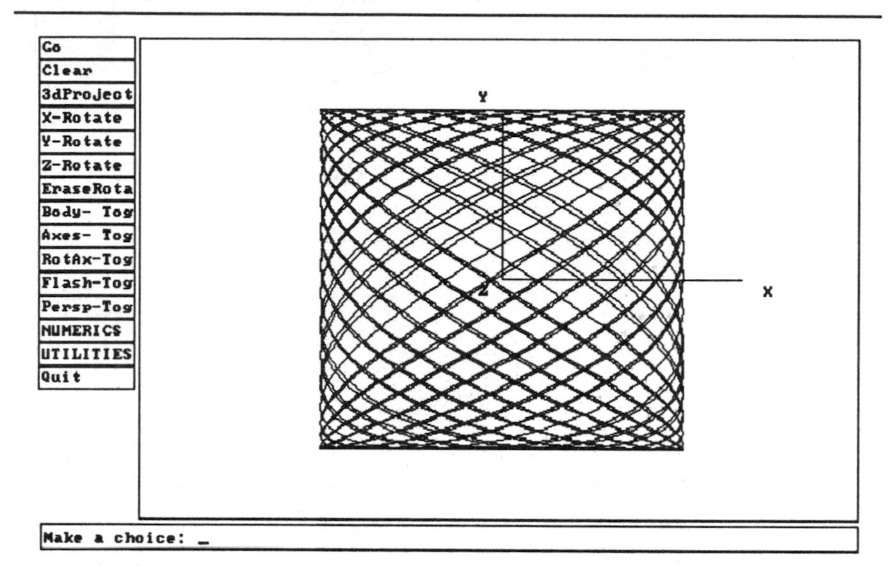

Figure 5.19. A Lissajous figure: orbit of a pair of linear harmonic oscillators (5.5).

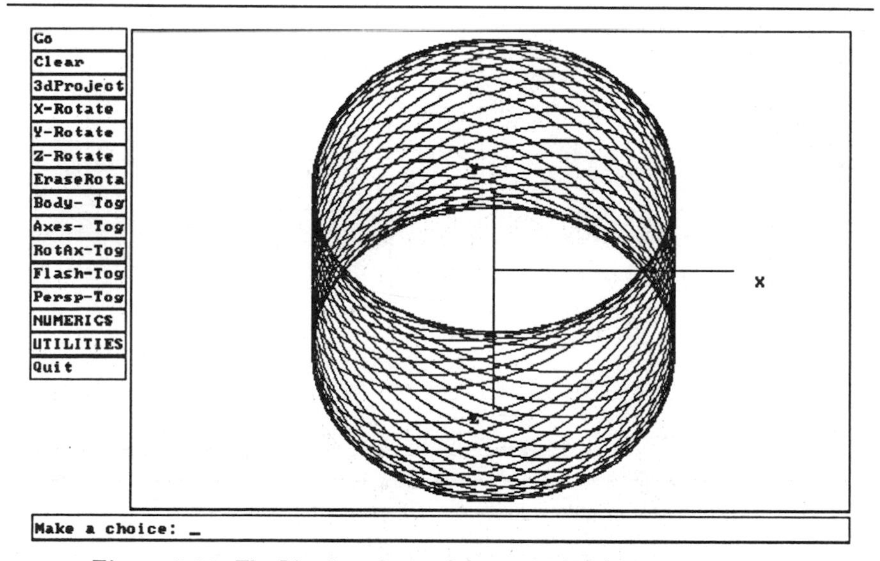

Figure 5.20. The Lissajous figure above rotated 60° about the x-axes.

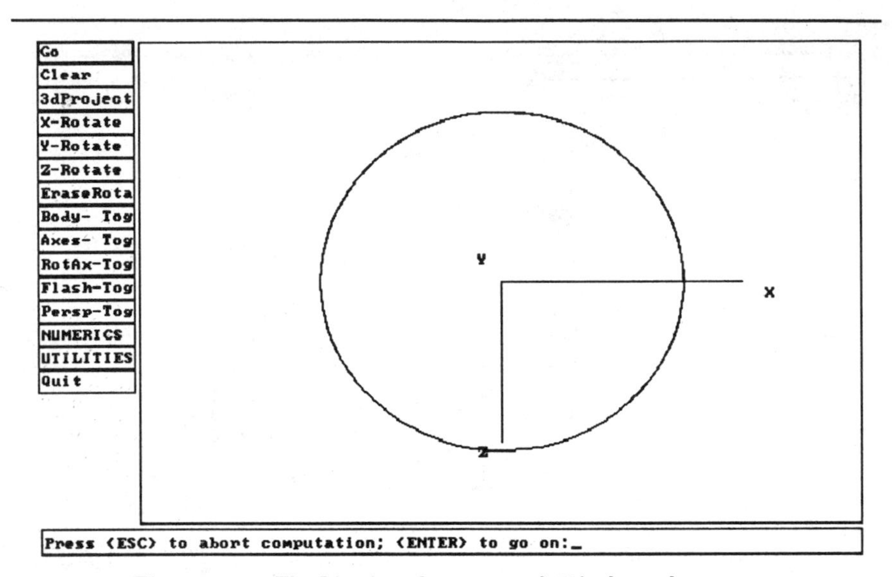

Figure 5.21. The Lissajous figure rotated 90° about the x-axes.

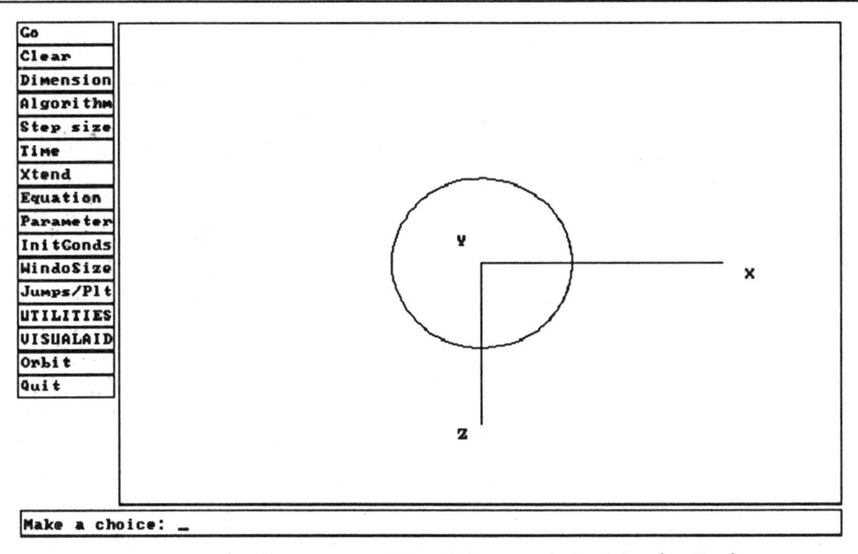

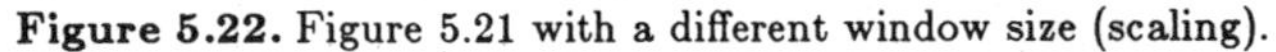

Figure 5.22. Figure 5.21 with a different window size (scaling).

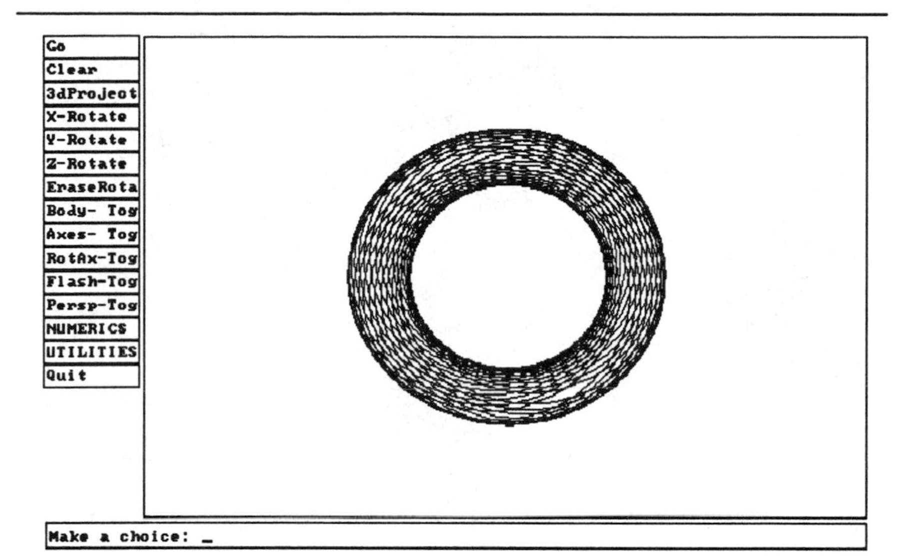

Figure 5.23. A perspective view of the Lissajous figure in Figure 5.22 from the view point $z = 30$.

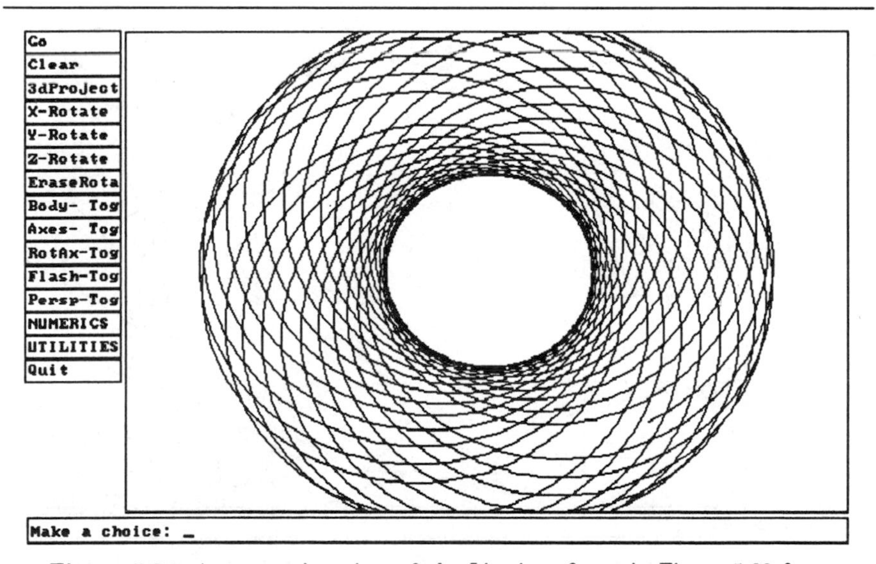

Figure 5.24. A perspective view of the Lissajous figure in Figure 5.22 from the view point $z = 15$.

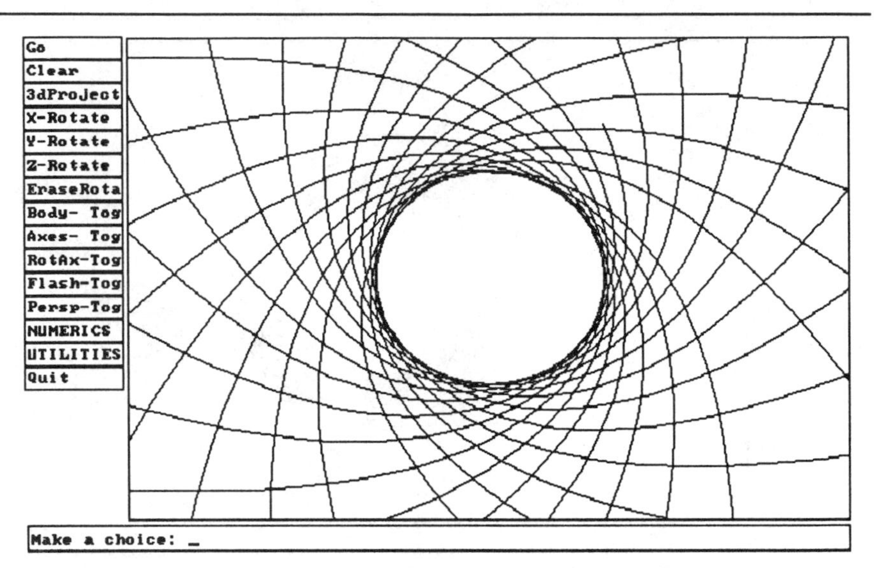

Figure 5.25. A perspective view of the Lissajous figure in Figure 5.22 from the view point $z = 5$.

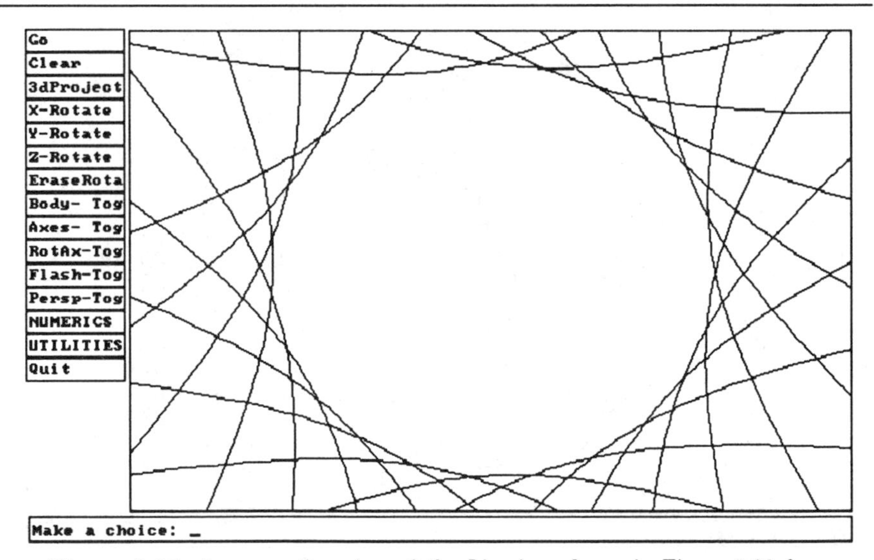

Figure 5.26. A perspective view of the Lissajous figure in Figure 5.22 from the view point $z = -5$.

Lesson 14. *Planar sections and Poincare maps.*

In this lesson, we will learn how to take sections and display Poincare maps of the cylinder, as introduced in the previous lesson. If you are continuing from Lesson 13, you may skip the first five steps below.

COMMAND:	RESPONSE/EXPLANATION:
d	*Dimension*: To change the current dimension.
4 <ret>	To make the current dimension 4. See Lesson 13.
t	*Time*: To change the current start and end times.
<ret>	To keep the start time the same (0.0).
65 <ret>	To make the end time 65.0.
u	*UTILITIES*: To bring up the UTILITIES menu.
t	*TwoViews*: To display two views.
c	*CutPlane:* To make *CutPlane* the top view. This is like the phase portrait view, except that the parts of all orbits on the positive side of a specified plane will be plotted in light blue, and those parts on the negative side of the plane will be plotted in dark blue. The equation of the plane in the screen coordinates is defined by $Ax+By+Cz+D=0$. The current (default) equation of the plane is $0x+1y+0z+0=0$.
m	*MapPoinc*: To make the bottom view *MapPoinc*, in which the Poincare map of the orbit on the plane specified above will be displayed.
g	*Go*: You should see *Figure 5.27* on the screen. In the *MapPoinc* view, the horizontal ($y=0$) plane is rotated about the center of the three-dimensional viewing box (90° about the x-axis) so that you are looking at it straight on. The default direction is positive (P); that is, only the points of intersection while the orbit is crossing the plane from the negative side to the positive will be plotted.
c	*Clear*: To clear the screen.
m	*MapPoinc*: To change the equation of the plane used in the *CutPlane* and *MapPoinc* views.
<ret>	To keep $A=0$.
0 <ret>	To set $B=0$.
1 <ret>	To set $C=1$.

<ret>	To keep $D = 0$.
<ret>	To keep direction the same (P). Now, in the *CutPlane* and *MapPoinc* views, the vertical plane through the origin ($z = 0$) will be used.
g	*Go*: You should see *Figure 5.28* on the screen. Due to the rotation of the plane, the Poincare map looks rotated.

You should try this lesson with two sets of initial conditions. Also, use *ValuPoinc* to see the values of the Poincare map. In case you try to use perspective, be warned that it works in the *CutPlane* view, but not in *MapPoinc*, because perspective does not preserve relative distances.

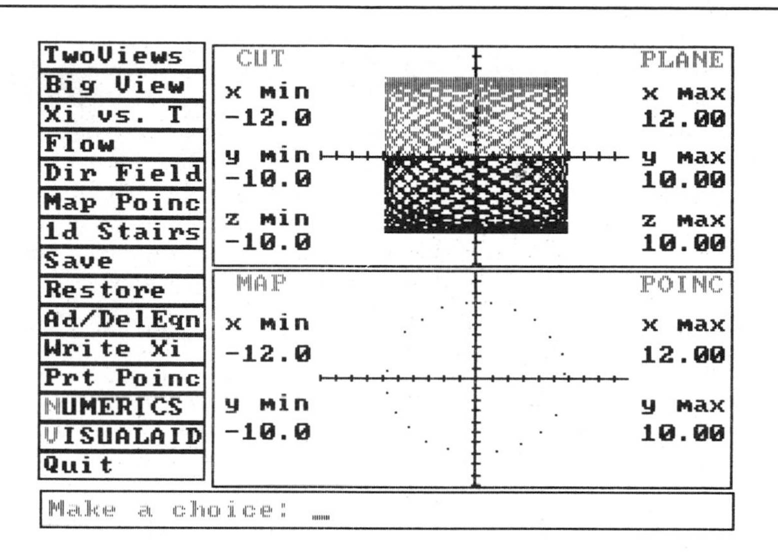

Figure 5.27. An orbit and its Poincare map of a pair of linear harmonic oscillators on the horizontal plane ($y = 0$).

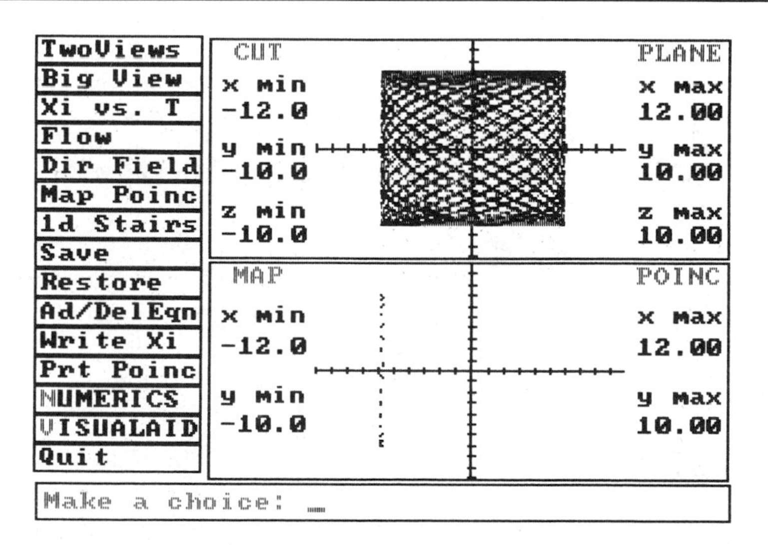

Figure 5.28. An orbit and its Poincare map of a pair of linear harmonic oscillators on the vertical plane ($z = 0$).

Lesson 15. *Grand Finale: demo.*

This is the final "lesson" with PHASER. It is a collection of screen images designed to demonstrate the basic capabilities of PHASER. You may recognize some of the pictures from the previous lessons. The others have been chosen to give you a synopsis of the fascinating dynamics of several representative equations from the library. We will refrain from presenting a mathematical discussion of these equations here; however, you may wish to consult the appropriate entries in Chapters 7 and 8 for details.

This demonstration transcript has been made using the *Save* entry on the *UTILITIES* menu. For further information, see Lesson 7 as well as the *Save* and *Restore* entries in section 6.2.

COMMAND:	RESPONSE/EXPLANATION:
u	*UTILITIES*: To bring up the UTILITIES menu.
r	*Restore*: To play back an existing transcript file, created using *Save*. You will be asked for a file name.
demo <ret>	To play back what is in the file *demo*. You will see the first picture in *demo*. Then, in the message line the prompt "Hit <SPACE> to go on; <RET> to exit" will appear. To see the next picture, hit the space bar. By repeating this process, you can play back all the pictures in *demo*. Finally, the familiar prompt "Make a choice: __" will signal the end of the transcript.

This concludes the practical training for PHASER. Now you are ready to use it either to further your mathematical education, or as a research tool.

Chapter 6

Reference Guide to Menus

This chapter contains detailed information about the entries of the three main menus (*NUMERICS, UTILITIES*, and *VISUALAID*) as well as the nine graphical views of PHASER (*PhasePort, Set up, Xi vs. T*, etc.); see *Figures 6.1-2.* You should first browse through the chapter to get acquainted with the capabilities of PHASER, and then return to it for reference purposes later. For a short synopsis, consult Appendix A.

To "make a choice" from a menu on the screen, just type in the first character of the name of the entry, in either lower-case or upper-case, but *do not* press the <return> (or the <enter>) key. The selection will be highlighted by a red box, and as soon as the command corresponding to the entry has been executed, the highlighting will be turned off.

In addition to the three main menus, various submenus come up when further choices are necessary. For instance, if you select the "Algorithm" entry by typing "a", then a submenu containing four entries, "Difference," "Euler," "ImpEuler," and "Runge-Kut," will appear; see *Figure 4.3.* To make *Euler*, for example, the current algorithm, just type "e". The submenu will be erased, and the NUMERICS menu will reappear. Selecting from a submenu is done in the same way as selecting from one of the main menus. The only exception is that when selecting from the submenu containing names of equations that appears after picking "Equation," you must type in not just the first letter but the full name of the equation of your choice, followed by <return>. As a general rule, you are expected to type the part of a menu or submenu entry that is in yellow or blue. To aid recognition, the first letters of the main menu names are written in blue.

NUMERICS	UTILITIES	VISUALAID
Go	Two Views	Go
Clear	Big View	Clear
Dimension	Xi vs. T	3dProject
Algorithm	Flow	X-Rotate
Step Size	DirField	Y-Rotate
Time	MapPoinc	Z-Rotate
Xtend	1d Stairs	EraseRota
Equation	Save	Body-Tog
Parameter	Restore	Axes-Tog
InitConds	Ad/DelEqn	RotAx-Tog
WindoSize	Write Xi	Flash-Tog
Jumps/Plt	PrtPoinc	Persp-Tog
UTILITIES	NUMERICS	NUMERICS
VISUALAID	VISUALAID	UTILITIES
Quit	Quit	Quit

Figure 6.1. The entries of the three main menus of PHASER: NUMERICS (left), UTILITIES (center), and VISUALAID (right).

If you select a menu entry that requires user input, you will get a prompt in the message line, showing in parentheses the current value of the expected input. For example, if you choose the "Dimension" entry and the current dimension is 2, you will be prompted with "Enter dimension (2) : __ ". If you want the current dimension to remain 2, just hit <return>. Otherwise, type in a different integer of your choice, followed by <return>. The general rule is that after making a choice from a menu, hit <return> if you wish to retain the current setting of the entry; otherwise, type in your input, followed by <return>.

6.1. The NUMERICS Menu

This menu provides functions for selecting options related mainly to numerical computations: step size, time interval, algorithm, etc. The entry *WindoSize*, for specifying the boundaries of the viewing box, is also placed on this menu because of its frequent use. When PHASER first starts up, the menu that appears on the screen is the NUMERICS menu.

Go

To commence computations using current settings. Results are automatically displayed inside the views that are on the screen. In the message line, a growing bar monitoring the progress of the computations appears. **The bar begins at time zero, and moves until the *End*-time, changing color from dark blue to light blue at *Start*-time when the plotting starts. If the computation time is extended using *Xtend*, the bar is rescaled from time zero to the new *End*-time.**

After typing *Go*, computations can be halted temporarily by pressing <ESC>, the escape key on the upper left corner of the keyboard. To continue computations, press <return>; to abort, press <ESC> again, and you will be returned to the current menu.

To prevent PHASER from "crashing," computations are stopped when the absolute values of numbers get too large (danger of *over-flow*), and the following message is displayed: "IC ... is going out of bounds at time ...". In the case of multiple initial conditions, computations are continued for the remaining ones.

Clear

To clear contents of views on the screen.

Dimension

To change the current dimension, which is the number of first-order equations in the current system. Enter an integer between 1 and 9, followed by <return>.

If the dimension is changed, all previous rotations are erased and the perspective option is set to the default value "no perspective".

The default dimension is 2.

Algorithm

To change the algorithm used to approximate solutions. When activated, a submenu containing the following four choices comes up:

- *Difference*: For iterating difference equations, or maps.
- *Euler*: The algorithm of Euler for differential equations.
- *ImpEuler*: Improved Euler, or Heun's method, for differential equations.
- *Runge-Kut*: Fourth-order Runge-Kutta for differential equations.

As soon as one of these algorithms is chosen from the submenu, the NUMERICS menu reappears.

The default algorithm is *Runge-Kutta*.

Step Size

To specify the step size to be used with the current algorithm. Enter the desired step size between -1 and 1, except 0, and press <return>. When the current algorithm is *Difference*, the step size is automatically set to 1. If you change the algorithm from *Difference* to one of those for differential equations, the step size is reset to the default value 0.1.

A negative step size can be used to run orbits backward, but make sure that signs of *Step size, Start-* and *End-*times are compatible, as explained below. Their consistency is checked after *Go* is pressed. If they are inconsistent, then an appropriate error message is displayed.

Time

To specify the *Start* and *End* of the time interval during which *plotting* is to be done. Enter the desired *Start*-time, followed by <return>; similarly for *End*-time. Solutions are always computed from time zero, but not plotted until *Start*-time. *Start* and *End* should either be both nonnegative or both nonpositive, with the absolute value of *Start* less than that of *End*-time.

Solutions of differential equations can be run backward by setting both the *Start-* and *End-*times and the *Step size* negative. This is an important technique for locating "sources" in phase portraits.

The default values are *Start* = 0.0 and *End* = 15.0.

Xtend

To extend the *End*-time of computations so as to continue solutions. Enter the new *End*-time, followed by <return>. The new *End*-time must be greater in absolute value than the old one. Solutions are automatically continued from the old *End*-time to the new *End*-time.

To prevent meaningless results, *Xtend* works only after *Go* or *Xtend*. Before extending again, changing the *Step Size*, *Clear*, and switching of menus are permitted. When exploring new equations, it is advisable to start with a small *Time* interval, and then *Xtend* if something interesting appears about to happen.

Equation

To change the current equation under study. A submenu of equation names appears. Type in the full name of the desired equation exactly as it appears on the submenu, followed by <return>. After a new equation is picked, the NUMERICS menu reappears.

Equations in the library are sorted by their dimension and also by their type -- that is, whether they are differential or difference equations. The difference/differential type is determined by *Algorithm*: *Difference* for difference equations (maps), and the remaining three algorithms, *Euler, ImpEuler*, and *Runge-Kut*, for differential equations. The submenu that comes up, therefore, contains only the equations of the current dimension and type. If there are more than nineteen equations in a category, the last entry of the submenu is *MORE*. To go to the next page of the submenu of equation names, type *MORE*, followed by <return>. The pages of the submenu cannot be "flipped backwards".

User-supplied equations (entered with the help of the *Ad/DelEqn* entry of the UTILITIES menu) are automatically added to the appropriate submenu of equations, and as far as the user is concerned, they become indistinguishable from the permanent library of PHASER.

For each dimension and algorithm type, there is a different default equation. The default equation is *linear2d* when the program is first started. For more information on equations, see the *Equation* entry in section 6.4, Chapters 7 and 8, and Appendix B. Lesson 8 of Chapter 5 describes how to browse through the equations stored in the library of PHASER.

Parameter

To change parameters in the definition of the current equation. A submenu of parameters, consisting of single letters, appears. Type the letter of the parameter to be changed, and then enter its new value followed by <return>; to keep the current value, simply hit <return>. To return to the NUMERICS menu, hit <return> again.

Up to the first six parameters and their current values in the current equation are displayed in the small version of the *Set up* view. If there are more parameters (up to twelve) in the equation, they can all be seen in the enlarged version of the same view. Because of the limited screen area, displays of parameter values are truncated to three digits. However, more digits of all parameters can be seen by picking them from the submenu of parameters and looking at the number in parentheses in the message line.

Note: Changing parameters interactively is an invaluable tool in exploration of bifurcations.

InitConds

To specify initial conditions. Enter the initial value of each variable xi, followed by <return>. After one set of initial conditions has been specified, you will be asked if you want to enter another set. Type **y** for yes, otherwise type **n** or just hit <return>.

A maximum of six sets of initial conditions can be entered at one time; their corresponding solutions will be plotted simultaneously in the *PhasePort*, *Xi vs. T*, *CutPlane*, and *MapPoinc* views. In *PhasePort*, *Xi vs. T*, and *MapPoinc*, the first solution is plotted in light blue, the second in red, the third in green, the fourth in white, and so on. In the *CutPlane* view, all solutions are plotted in light blue on the positive side of the plane and in dark blue on the negative side. When simultaneity is not essential, as many initial conditions as desired can be displayed in the same view, as long as the view is not cleared.

The default initial conditions are xi = 5.0, for all i.

WindoSize

To specify x-min, x-max, y-min, and y-max in dimension two, and in addition z-min and z-max in dimension three or higher. Enter each number, followed by <return>.

By convention the x- and y-axes are the horizontal and the vertical axes on the screen, and the positive z-axis perpendicular to the screen, coming towards the viewer. We will refer to these axes as the *screen axes*. Marked axes are placed not at the origin, but rather at the center of the plotting area determined by x-min, x-max, y-min, and y-max. After rotations, projections, etc., final pictures are clipped by the boundaries of this two-dimensional plotting area. The boundaries determined by *WindoSize* are used by the *PhasePort*, *CutPlane*, *MapPoinc*, and *1d Stairs* views. The *Xi vs. T* view has its own scaling that can be adjusted by activating the *Xi vs. T* entry on the *UTILITIES* menu. The current settings of boundaries are displayed on the sides of the small versions of these views. Entries of *WindoSize* must be less than 10^6 in absolute value. The defaults are

$x - min$ = -14.0, $x - max$ = 14.0, $y - min$ = -10.0, $y - max$ = 10.0, $z - min$ = -10.0, and $z - max$ = 10.0.

Note: A certain amount of experimentation may be required to find an appropriate *WindoSize* when studying an unfamiliar equation. Use the *Last Xi* view (see section 6.4) to get an idea about the size of the numbers you are trying to plot. Also, by adjusting the *WindoSize*, you can blow up those regions where the "action" is.

Jumps/Plt

To specify the number of jumps (steps) to be taken between two consecutive plotted points. Enter a positive integer, followed by <return>.

This feature is useful when a small step size is necessary for numerical accuracy, but you may prefer to plot solutions at larger intervals, say every fifth step, to speed up graphical display. It can also be used to detect periodic orbits of difference equations. Be careful if you *Xtend* when *Jumps/Plt* is not 1. For example, you may inadvertently switch from odd to even points.

The default value is 1; that is, solutions are plotted at every step.

UTILITIES To bring up the UTILITIES menu.

VISUALAID To bring up the VISUALAID menu.

Quit To exit PHASER and return to the operating system (DOS Version 2.0 or higher).

6.2. The UTILITIES Menu

This section contains descriptions of the entries of the UTILITIES menu. The menu provides two basic groups of functions. The first group is for selecting various combinations of views. The second group is designed to facilitate the saving of pictures and/or lists of numerical output, and the entering of user-defined equations. Although the UTILITIES menu does not contain the *Go* and *Clear* commands, you can still use them. This is not true for any other menu entry.

TwoViews To choose two different views for display in the viewing area. The UTILITIES menu will be erased, and a submenu containing the names of nine possible views will be drawn (see section 6.4 for their descriptions). When prompted to pick a top view, type the first letter of the name of the desired view; similarly for the bottom view. If you wish to keep a view that is already on the screen, just hit <return>.

Big View To choose an enlarged version of one view for display in the entire viewing area. The UTILITIES menu will be erased, and a submenu containing the names of nine possible views will appear (see section 6.4 for their descriptions). Enter the first letter of the name of the view you wish to enlarge.

Xi vs. T To specify the subscript i of the variable Xi to be plotted in the *Xi vs. T* view, and the minimum and maximum values of Xi. Enter values, followed by <return>. Notice that these boundaries are independent of the numbers specified by *WindoSize.*

The defaults are: i = 1, Xi-min = -14.0, and Xi-max = 14.0.

Flow To plot the flow of the current equation in the phase portrait view. (This feature is available for dimension two only!) Enter an integer n between 2 and 30 for the grid size, followed by <return>. An $n \times n$ grid of equally spaced green dots will appear. Then, one at a time, the orbits emanating from each green dot will be drawn using current settings, like *Step size, Time*, etc. Computations can be halted by pressing the escape key <ESC>; to resume the procedure, press <ret>; to abort, press <ESC> again.

Note: Because this is a slow process, start with a small grid size and a short time.

DirField To draw the direction field of the current equation in the phase portrait view. (This feature is available for dimension two only)! Enter an integer n between 2 and 40 for grid size, followed by <return>. An $n \times n$ grid of equally spaced green dots will appear. Then at each dot a normalized white direction line segment will be drawn. For an example, see Lesson 3 in Chapter 5.

Note: Because of its display speed, the direction field is a good exploratory tool to see if anything exciting is likely to happen.

MapPoinc To specify the values of A, B, C, and D in the equation of the plane $Ax + By + Cz + D = 0$, as well as a negative or positive (N/P) direction. The direction is taken to be positive when an orbit crosses the plane from the negative side $(Ax + By + Cz + D < 0)$ to the positive side. The equation of the plane is always defined in the screen coordinates, and the plane is not effected by rotations.

This plane is used in two different ways:

- To compute a planar Poincare map to be displayed in the *MapPoinc* view (see section 6.4).
- In the *CutPlane* view, to display in light blue the parts of solution curves on the positive side of the plane, and in dark blue those on the negative side (see section 6.4.).

Default values are A = 0, B = 1, C= 0, and D = 0, and the default direction is P.

1d Stairs

To specify the power or the iterate (see section 3.1) of the current one-dimensional difference equation (map) to be displayed in the *1d Stairs* view. Enter a positive integer, followed by <return>.

This feature is useful for studying periodic orbits. For example, if the number is 2, then the intersections of the 45^0 line with the second power of the map give periodic orbits of period 2. For an example, see Lesson 11 in Chapter 5.

The default value is 1.

Save

To save the current window configuration and all settings in a user-specified file. Enter the file name, using a maximum of eight characters (any legal DOS file name), followed by <return>. If the file already exists, you will be asked to **a**ppend or to **o**verwrite (a/o).
When you overwrite, the old contents of the file are erased and a new transcript is started.

At all times, PHASER keeps track of every bit of information that is necessary to recreate the current picture on the screen. When you *Save*, all the current settings of the step size, the initial conditions, the equation, etc., and all the current views in the viewing area are recorded in a user-specified file on the disk. When you *Restore* from an existing file (see below), all this information is reloaded into PHASER and the picture is computed, as if the menu entry *Go* were activated. While changing parameter values or rotations, for example, you can keep appending to the same file so as to make a slow film of a certain dynamical behavior. See Lesson 7 and also Lesson 15, the *Grand Finale*, in Chapter 5 for more information on these rather unique capabilities of PHASER.

Notes: The direction field and flow cannot be saved. If you no longer need a saved transcript file, you should remove it from the disk while in the DOS operating system by issuing the command **erase <filename> <ret>**.

Restore

To play back a transcript from an existing file created using *Save*. Enter file name, followed by <return>. All information necessary to recreate one picture is loaded into PHASER, and *Go* is activated. To see the next picture, press the space bar; otherwise, hit <return> to terminate the transcript. At the end of a playback, all settings are the ones associated with the last picture.

Save and *Restore* are extremely powerful facilities enabling instructors to prepare "electronic lectures" for classroom demonstrations. They are also useful for saving the current environment in a form that can be reloaded into PHASER at a later time for further work.

Ad/DelEqn

To add or delete user-defined equations to/from the library of PHASER. This is one of the most powerful facilities of our simulator, allowing each user to modify the library of equations according to her/his needs. The process is somewhat complicated at first, but from the user's point of view, once entered, new equations can be accessed and simulated just like the other library equations.

Adding an Equation: Before adding your own equations to PHASER, a certain amount of preparation is required. First, if your equation is not already a first-order system (and autonomous, in the case of differential equations) you should convert it into this form using the techniques of Chapters 1 and 3. Next, you must decide on the name of your equation, and identify its parameters, constants, etc., as explained below.

We will use the *logistic* equation,

$$x_1^{k+1} = a x_1^k (1.0 - x_1^k),$$

to illustrate the process. For another example, see Lesson 9 in Chapter 5. To enter an equation, you must complete the following seven steps:

Step 1. After activating the *Ad/DelEqn* entry of the UTILITIES menu, you will be asked: "Want to add or delete eqn? (a/d)". To add a new equation, type **a <ret>**.

In case you have already reached the maximum number of allowable equations (about forty), you will get the message "No space left for new equation!"; "Delete an old one to make space." At this point, just hit <ret> and delete some old equations, as explained below, before entering your new one.

Step 2. You will be asked to "Enter equation name (MAX 8)". If you do not wish to continue, just hit <ret> and you will be returned to the UTILITIES menu. Otherwise, type in a maximum of eight letters and/or digits as the name of the new equation and then hit <ret>.

The new name must be different from any already existing equation name. So, let us call the logistic equation *logis* and type in **logis <ret>**.

Step 3. You will be asked whether your equation is "0 = Difference, 1 = Differential (0/1)". If the new equation is a difference equation, type 0; otherwise, type 1. Then hit <ret>. For *logis* you should type **0 <ret>**.

Step 4. You will be asked to "Enter dimension (1 - 9)". Type in an integer between 1 and 9 and hit <ret> (the maximum dimension is 9). Since *logis* is one-dimensional, type **1 <ret>**.

Step 5. You will be asked to "Enter number of parameters (0 - 12)" in your equation. Type in an integer between 0 and 12, followed by <ret> (0 is for none, and the maximum number of parameters allowed is 12). For *logis* you should enter **1 <ret>**.

Next, you will be asked to enter the name and the default value of each parameter by the following two prompts: "Enter name of param <number> (1 letter)" and "Enter value of <parameter>". For *logis* you should enter **a <ret>** and **2.7 <ret>**. These default values can be changed during simulations by using the *Parameter* entry of the NUMERICS menu.

Step 6. You will be asked to enter "Number of constants (0 - 9)" in your equation. Type in an integer between 0 and 9, followed by <ret> (0 is for none, and the maximum number of constants allowed is 9). *Constants* are simply all the real numbers appearing in your equation as factors, terms, exponents, etc. Do not confuse constants with parameters. For *logis* you should enter **1 <ret>** because there is only one constant. Constants will automatically be named c1, c2, ... up to the number of constants declared above.

Next, for each constant you will be asked to "Enter value of c<i>". In *logis* you should enter **1 <ret>** for the value of c1. Declaring real numbers in equations as constants is for the convenience of entering your equation into PHASER. Once a new equation is entered, all constants will be replaced by their values, and their names can be safely forgotten.

Step 7. At this point you will be asked to enter your equation "x1' = ". Type in the formula for x1', terminated by <ret>. In the formula, a maximum of seventy nine characters (including blank spaces) is allowed, but only the first thirty five of them will be visible in the message line. Therefore, you should be extremely careful when typing a long formula. You will be prompted to enter a formula for each equation in your new system up to the dimension of the system. For *logis* you should type **a * x1 * (c1 - x1) <ret>**.

Here is the list of all the allowable symbols you may use in your formula:

Variables: **x1**, **x2**, ... , up to the dimension.

Parameters: Letters defined in step 5.

Constants: **c1**, **c2**, ... , up to the number defined in step 6.

Variables, parameters, and constants are called *operands*.

Binary operators: **+** for addition, **-** for subtraction, **/** for division, ***** for multiplication, and **^** for power (example: $a\,\hat{}\,b = a^b$).

Unary operators: **-** for unary minus, **sin()** for sine, **cos()** for cosine, **exp()** for exponentiation, **log()** for natural logarithm, **abs()** for absolute value, and **flr()** for floor function, which is the largest integer less than the number inside the parentheses. A common use of the floor function is the following identity:

$$a \ mod \ b \equiv a - b \ \mathbf{flr}(a/b) \ .$$

Left and right parentheses: **(** and **)**.

After you enter a formula, it will be checked for possible syntax errors. You will be returned to the UTILITIES menu if the formulas are acceptable. Otherwise, you will get error messages, such as "Operator followed by operator," "Imbalanced parentheses," etc., and you will be given a chance to reenter your formula. After entering a new equation, make sure to verify its text by using the *Equation* view.

Deleting an Equation: It is quite easy to delete a user-defined equation from the library. However, the equations in the fixed library cannot be deleted. PHASER has enough space for about forty user-defined equations. Therefore, you should periodically delete the old equations that are no longer needed.

When you activate the *Ad/DelEqn* entry of the UTILITIES menu, the prompt "Want to add or delete eqn? (a/d)" will appear. To delete a user-defined equation, type **d**, followed by **<ret>**. Next you will be asked to "Enter equation name". Type in the name of the equation you wish to delete, followed by <ret>; the equation will be deleted.

If you type in the name of an equation from the fixed library, you will get the error message "Cannot delete from fixed library!". In case you type in a name that does not exist, you will be warned that "There is no such equation!". In both instances, you will be given a chance to try again. If you change your mind and wish to return to the UTILITIES menu, just press <ret>.

Write Xi

To write (unrotated) computed values of all variables xi in a user-specified file. Type in the desired file name, containing a maximum of eight characters, followed by <return>. Using current settings, solutions will be computed as if *Go* were picked, and the results will be written into the file rather than displayed graphically. After *Quit*ting PHASER, the file can be examined by typing the DOS command **type <filename>**. For a sample output, see *Figure 2.1*.

PrtPoinc

To print, or write, (unrotated) computed values of a Poincare map in a user-specified file. Type in the desired file name, containing a maximum of eight characters, followed by <return>. Using current settings, the Poincare map will be computed as if *Go* were activated, and the results will be written into the file rather than displayed graphically. The *Time* recorded in the file is the time at which the solution crosses the plane in the specified direction. After *Quit*ting PHASER, the file can be examined by typing the DOS command **type <filename>**.

NUMERICS To bring up the NUMERICS menu.

VISUALAID To bring up the VISUALAID menu.

Quit To exit PHASER and return to the operating system (DOS version 2.0 or higher).

6.3. The VISUALAID Menu

This section contains detailed descriptions of the entries of the VISUALAID menu as well as a brief discussion of three-dimensional graphics. The menu provides functions related to three-dimensional graphical aids such as projections, rotations, perspective, etc. Viewing a three-dimensional object on a two-dimensional screen can be a complicated process. Here we will present a short overview of selected concepts from 3D computer graphics, as they are implemented in PHASER. For a more comprehensive discussion of this subject, you should consult Foley & Van Dam [1983].

In 3D viewing we first specify boundaries of a three-dimensional view box. This is done using the *WindoSize* entry on the NUMERICS menu. Then we place the *screen axes* at the geometric center of this box as follows: the x-axis is horizontal, the y-axis is vertical, and the positive z-axis is perpendicular to the screen, coming towards the viewer. The most common way of projecting an image in 3D onto the screen -- that is, the (x, y)-plane -- is done by simply dropping the last coordinate, which is called the *orthographic projection*. Then the projected image is clipped by the boundaries of the two-dimensional viewing area determined by the x-min, x-max, y-min, and y-max of the view box.

To gain a good feel for the shape of a three-dimensional object on a two-dimensional screen, it is often both helpful and necessary to perform rotations in 3D. To accomplish this interactively, three fundamental rotations around the fixed x, y, and z screen axes are provided. (PHASER adheres to the so-called "right-hand rule" in all rotations.) Since rotations are handled cumulatively, any 3D rotation can be achieved through a sequence of these fundamental rotations. The cumulative effect of rotations can be seen by displaying the rotated axes, using *RotAx-Tog* on the VISUALAID menu. Once the desired rotated configuration is achieved, images can be drawn by activating *Go*.

It is important to keep in mind that rotations using *X-Rotate*, *Y-Rotate*, and *Z-Rotate* on the VISUALAID menu are performed with respect to the original screen axes. However, these fundamental rotations can be turned into *body rotations* (in other words, performed with respect to the rotated axes) by toggling on the body rotation option, using the *Body-Tog* command. When PHASER starts up, this option is off. Once the body rotation option is turned on, all the fundamental rotations are performed as body rotations until this feature is turned off.

In addition to orthographic projection, PHASER provides one-point *perspective projection* for viewing three-dimensional objects on a 2D screen. The perspective projection creates a more realistic visual effect: the parts of an object closer to the point of view look bigger, thus giving a sense of depth. It is not without drawbacks, however, because distances cannot be determined from the projection, and parallel lines may not look parallel. For more details on perspective projection, see the *Persp-Tog* entry on the VISUALAID menu.

Some experimentation may be necessary until you get used to certain subtleties of three-dimensional computer graphics. For example, suppose that x-min = y-min = -10, z-min = -20, and x-max = y-max = 10, z-max = 20. The point in 3D with coordinates (0, 0, 30) is plotted at the center of the viewing area on the screen. If the point is rotated 90° about the x-axis, then the new coordinates of the point are (0, -30, 0). Under orthographic projection this point would be plotted at (0, -30) on the screen. However, this falls outside the two-dimensional viewing area determined by the boundaries of x and y. Therefore, the rotated point will not be visible on the screen.

Go

To compute and display. See the description in the NUMERICS menu.

Clear

To clear contents of views.

3dProject

To project onto a three-dimensional subspace from higher dimensions. You will be asked to specify subscripts of three dependent variables to be assigned to the x, y, and z-axes for plotting. For example, to display x1, x2, and x4 on the x, y, and z-axes, respectively, type 1 <return>, 2 <return>, and 4 <return>.

Default assignments are x1, x2, and x3 onto the x-, y-, and z-axes, respectively.

X-Rotate

To rotate images about the x-axis; that is, during rotation, the x-axis will remain fixed. Enter the amount of the desired rotation in degrees, followed by <return>. In all rotations the so-called "right-hand rule" is used.

Since rotations are handled cumulatively, any 3D rotation can be achieved through a sequence of *X-Rotate, Y-Rotate*, and *Z-Rotate* commands. The cumulative effect of rotations can be seen by toggling on the rotated axes, using the *RotAx-Tog* entry on the VISUALAID menu. Once the desired rotated configuration is achieved, images can be drawn by activating *Go*. You may want to clear the screen between rotations.

Y-Rotate To rotate images about the y-axis. Enter amount of rotation in degrees, followed by <return>.

Z-Rotate To rotate images about the z-axis. Enter amount of rotation in degrees, followed by <return>.

EraseRota To erase all previous rotations. Axes will be set to the original unrotated configuration. It is difficult to undo a long sequence of rotations because they are cumulative as well as noncommutative. To start afresh after several rotations, it is best to use *EraseRota* rather than to undo them one at a time. If the dimension or equation is changed, all previous rotations are erased automatically.

Body-Tog To toggle body rotations on and off. By default they are off. If turned on, all rotations are performed as body rotations until turned off. In other words, if *Body-Tog* is on, *X-Rotate* will rotate an object about the rotated x-axis, not about the original screen x-axis. When this feature is on, the word *Tog* will be highlighted in red.

Axes-Tog To toggle on and off the marked axes in the *PhasePort*, *CutPlane*, and *MapPoinc* views. By default, they are on. Axes are placed at the center of the plotting area, not at the absolute origin. While doing rotations, turn this option off, and turn on the rotated axes, as explained below.

RotAx-Tog To toggle rotated axes, also called body axes, on and off. By default, they are off. While doing rotations, turn this option on to see the rotated configuration. When rotated axes are on, the word *Tog* will be highlighted in red.

Flash-Tog To toggle on and off a flashing marker at the current point being plotted. By default, it is off. The flasher is a very helpful visual aid in certain situations: tracing periodic orbits, running two solutions simultaneously, and, of course, interpreting cluttered pictures. When the flasher is on, the word *Tog* will be highlighted in red.

Persp-Tog To project from 3D to 2D using *one-point perspective projection.* You will be asked to provide the distance of the *viewing point* on the screen z-axis. Enter a number greater than $z\text{-}min$, followed by <return>. For no perspective, enter **n** <return>, which is the default value.

The plane of projection is the $z = z\text{-}min$ plane, and the viewing point is on the z-axis. The parts of an object whose z-coordinates are greater than that of the viewing point (behind the eye) and less than $z\text{-}min$ (behind the plane of projection) are clipped. Then the remaining parts are projected onto the plane $z = z\text{-}min$ using one-point perspective projection. Keep in mind that parts of objects closer to the point of projection will look bigger. You can "walk through" an object by varying either the point of view or the plane of projection (changing $z\text{-}min$). If there is any perspective, the word *Tog* will be highlighted in red. See Lesson 12 in Chapter 5 for a detailed example.

Note: Perspective is not in effect for the *MapPoinc* view.

NUMERICS To bring up the NUMERICS menu.

UTILITIES To bring up the UTILITIES menu.

Quit To exit PHASER and return to the operating system (DOS version 2.0 or higher).

6.4. The Graphical VIEWS

This section contains detailed descriptions of the nine graphical views of PHASER. By using the *Two Views* and *Big View* entries on the UTILITIES menu, any two small or one enlarged version of the nine views can be displayed. When PHASER starts up, *PhasePort* is the top view and *Set up* the bottom view. Using this default configuration, you should first adjust all numerical settings, then put up different views as desired.

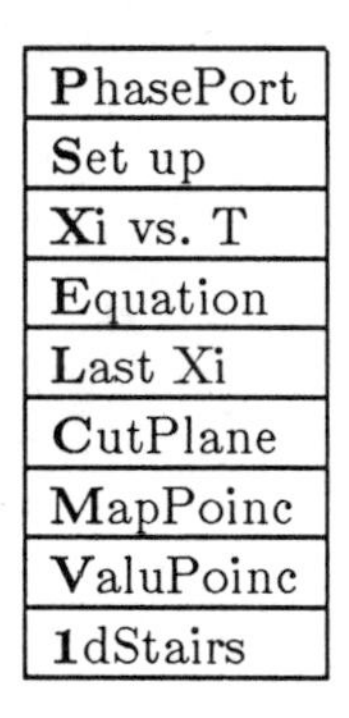

Figure 2.6. Nine graphical views of PHASER.

PhasePort To display solutions, flows, and direction fields. In the small version, the boundaries of the plotting region are displayed on the sides.

Set up To display the current settings of "all things". If any setting is changed, it is updated immediately. Here is a line-by-line account of this view:

• *Line 1*: Name of equation, dimension, and algorithm. These can be changed by activating the *Equation*, *Dimension*, or *Algorithm* entries on the NUMERICS menu. Defaults are *linear2d*, 2, and *Runge-Kutta*.

• *Line 2*: Values of any parameters in the definition of the current equation. In the small version, only the first six parameters are displayed. In the enlarged version all (up to twelve) are visible. Their values can be changed by activating the *Parameter* entry on the NUMERICS menu.

- *Line 3*: *Start* and *End* of time interval for display. They can be changed by activating the *Time* entry on the NUMERICS menu. Defaults are 0.0 and 15.0.
- *Line 4*: Step size and jumps (steps) between two consecutive plotted points. These can be changed by activating the *Step size* and *Jumps/Plt* entries on the NUMERICS menu. Defaults are 0.1 and 1.
- *Line 5*: Assignment of variables to the x-, y-, and z-axes (the rotated axes, if any rotations are performed). They can be changed by activating the *3dProject* entry on the VISUALAID menu. Defaults are x1, x2, and x3.
- *Line 6*: The equation, $Ax + By + Cz + D = 0$, of a plane, and a negative or positive direction (N/P). The values of the coefficients A, B, C, and D in the equation and the direction can be changed by activating the *MapPoinc* entry on the UTILITIES menu. Defaults are $A = 0$, $B = 1$, $C = 0$, $D = 0$, and P.
- *Line 7 and more*: One set of initial conditions on each line (only the first four variables are visible). Initial conditions can be changed by activating the *InitConds* entry of the NUMERICS menu. In the small version, only the first two sets are visible. In the enlarged version, all six possible sets are displayed.

Xi vs. T

To display the graph of the i-th variable Xi vs. time. The time axis runs from *Start* to *End*-time. If the *End*-time is extended using *Xtend*, then the time axis runs from the old end-time to the new extended time. This view has its own scaling, which is displayed on the side of the small version. The variable to be plotted and the corresponding scaling are specified using the *Xi vs. T* entry of the UTILITIES menu.

Defaults are $i = 1$, $xi - min = -14$ and $xi - max = 14$.

Equation

To display the text of the current equation. When this view is in effect, the equation is displayed automatically.

If you are studying a new equation or have entered one of your own, make sure to verify its text using this view.

Last Xi

To display the last several (rotated) values of the three projected variables. To see the values, activate *Go*. In the small/big version, the last eight/twelve lines of data are displayed. In the case of multiple initial conditions, solutions are numbered and their values are listed consecutively at each iteration. This view is useful for looking at values of variables in order to get an idea of their magnitudes, so that the window size can be adjusted accordingly. To get a printout of all values of variables, use the *Write Xi* entry on the UTILITIES menu.

CutPlane

To display solutions in two colors on different sides of the plane specified by the *MapPoinc* entry on the UTILITIES menu. To see the solutions, activate *Go*. Solutions are plotted in light blue while on the positive side of the plane and in dark blue while on the negative side. The same color code is used as well for all solutions in the case of multiple initial conditions. Notice that this color coding differs from the one used in the phase portrait view, where each solution is colored differently.
This view can be used to create a sense of depth by varying the plane, say $z = constant$. It is best to use *CutPlane* simultaneously with the *MapPoinc* view.

MapPoinc

To display a planar Poincare map. The plane and the direction are specified using the *MapPoinc* entry of the UTILITIES menu. To see the Poincare map, activate *Go*. The plane is automatically rotated about the center of the three-dimensional viewing box determined by *WindoSize*, so that the plane is perpendicular to the screen z-axis. The boundaries of the rotated Poincare plane are determined by the x-min, x-max, y-min, and y-max values specified by the *WindoSize* command. In the case of multiple initial conditions, solutions are plotted in different colors.

The perspective command, *Perp-Tog*, is not implemented for the *MapPoinc* view because distances are not preserved under perspective projections.

The computation of "useful" Poincare maps usually involves a certain amount of experimentation with different cut planes and window sizes. When interpreting pictures, also keep in mind the automatic rotation of the plane. To find an appropriate window size, you may want to use the *ValuPoinc* view to get an idea about the magnitude of numbers. Once you find a good setting, you should use the *CutPlane* and *MapPoinc* views together while exploring the geometry of solutions in 3D and higher.

ValuPoinc

To display the last several (rotated) values of a Poincare map. To see the Poincare values, activate *Go*. In the small/big version, the last eight/twenty lines of data are visible. To write all values of Poincare maps onto the diskette, use the *PrtPoinc* command on the UTILITIES menu.

1d Stairs

To display a stair step diagram of 1d difference equations (maps). To see the stair step diagram, activate *Go*.

This view uses somewhat different conventions than the others. Axes are placed at the absolute origin (0, 0) -- unlike *PhasePort*, for example -- and the 45° line is drawn through the origin in red. Boundaries of the viewing area are determined by the x-min, x-max, y-min, and y-max set by *WindoSize*. When *Go* is activated, first the graph of the function (or its power, if it is specified using the *1dStairs* entry on the UTILITIES menu) is plotted, and then the stair step diagram corresponding to the first initial condition alone.

Note: This view is not synchronized with the *PhasePort* view. If they are both open, *1d Stairs* is computed first (no moving bar), then *PhasePort* in the usual way.

Part III

Library of Equations

Chapter 7
Differential Equations

This chapter is a catalogue of the differential equations stored in the permanent library of PHASER. A similar collection of difference equations is presented in the next chapter. In addition, Appendix B contains a compressed list of all the equations, for quick reference. If you are not certain how to access a specific equation in the library, consult the *Equation* entry in section 6.1, as well as Lesson 8 of Chapter 5.

There is an entry for each equation containing its formula, a brief description, and specific references to the mathematical literature. In the formulas the variables are denoted by x_i, and their derivatives by x_i'. Single letters in lower case are parameters, which can be changed interactively, as explained in the *Parameter* entry in section 6.1. The listed values of the parameters are their default values. Following each formula, there are explanatory remarks on the practical or theoretical importance of the equation. These brief, and at times over-simplified, remarks are meant to capture your attention and arouse your enthusiasm for further study. The selected references at the end of each equation should provide a point of entry into the literature, where more detailed information can be found.

For each dimension, the equations are ordered roughly by increasing level of difficulty. One quarter of the library consists of central examples for an undergraduate course in conjunction, say, with Boyce & DiPrima [1977] or Braun [1983]. Another quarter is designed for a graduate course at the level of Guckenheimer & Holmes [1983]. The rest of the library is necessarily a reflection of my own mathematical inclinations. If you do not see your favorite classroom example or research problem, do not be alarmed; new equations can easily be added to the library, as explained in the *Ad/DelEqn* entry in section 6.2, and Lesson 9 of Chapter 5.

At the end of each section there are collections of screen images depicting many interesting and important dynamical phenomena. To give you an idea of many of the essential choices (specific initial conditions, parameter values, etc.), in most figures the bottom view is *Setup*. Using these settings as a starting point, you should first try to recreate these pictures in order to get a feel for their *dynamics*. Afterwards, you can perform further numerical experiments for more in-depth study. Do not be discouraged, however, if you do not get the "right" picture immediately. Being an experimentalist, even on the computer, can be a time-consuming activity.

What is in an equation ?

PHASER has been designed with a diverse group of users in mind, and this consideration is reflected in the collection of equations in the library. However, the underlying theme of the library is the study of qualitative properties of differential and difference equations. The first stage in this pursuit is to determine the phase portrait of an equation by varying the initial conditions. In particular, one would like to know the limiting behavior of all the solutions of an equation as $t \to \pm\infty$; i.e., one would like to determine the "limit sets" of all the solutions of an equation. In general, these sets grow in complexity as the dimension of the equation increases. For example, in one dimension, a solution of a differential equation approaches an equilibrium point in forward or reverse time. In two dimensions, the possible limit sets in addition to equilibria are of two types: periodic orbits, and the collection of equilibrium points with orbits joining them. In three dimensions and higher, there are extremely complicated examples of limiting behavior, with little hope of a complete classification. A set that attracts nearby solutions in forward time, but is more complicated than an equilibrium point or a periodic orbit, is called a ***strange attractor***. Although there is no standard mathematical formulation of this concept, see Guckenheimer & Holmes [1983, p. 256] for an attempt to include many known examples exhibiting complicated dynamics. In the case of difference equations, the situation is still more complicated: even in one dimension, there are examples not merely with fixed points and periodic orbits, but also with strange attractors. All these phenomena, and more, can be seen in the equations stored in the library of PHASER.

The second stage in this qualitative study is to explore the possible changes in the phase portrait of an equation as the equation itself is varied. In applications, for example, many models contain changeable parameters. Even when this is not the case, it may be necessary to introduce parameters into a model so that by changing them the "robustness" of the system under small perturbations can be investigated. The study of qualitative changes (for example, variations in the number or the stability type of equilibria) in the phase portraits of dynamical systems as parameters are varied is called *bifurcation theory.* For given parameter values, a system is called *structurally stable* if small changes in the parameters do not change the qualitative properties of the phase portrait. Any other value of the parameters is called a *bifurcation value.* It is important to locate the bifurcation values, and to classify the possible phase portraits around them. Indeed, a substantial number of the examples in the library are designed to illustrate many of the "typical" bifurcations.

The ideal use of computers in dynamical systems is both to observe known dynamical phenomena and to discover new ones in specific examples. Many of the equations in the library possess complicated dynamics, which cannot be understood simply by plotting orbits at random, however. Therefore, before undertaking the study of an equation, you should master the precise mathematical formulation of the phenomena you hope to observe. Remarks and references at the end of each entry in the library are intended to facilitate this by pointing to sources where specific definitions, theorems, or more detailed information about a particular equation can be found. It is, of course, more difficult to suggest how to discover new phenomena. At the very least, you should keep in mind that theory and experimentation are mutually beneficial, especially when used iteratively.

An apology

I feel compelled to end this introduction to the library with an apology, or a "disclaimer," if you will. Frankly, I found it quite difficult to capture in merely a few sentences the mathematical or the practical essence of some famous equations. It would be easy to write a whole book about certain of the equations, and indeed there are such books. I hope that the loosely set down, and perhaps chatty, remarks below will encourage the novice to tinker with a few of the equations, thus arousing her/his enthusiasm for further study of their mathematical details. As for those of you who are more sophisticated readers, I beg your forgiveness and ask that you simply look at the formulas. This book, after all, is designed to be only a manual.

7.1. 1D Differential Equations

---

cubic1d *General one-dimensional cubic differential equation.*

Type: 1D Differential.

Parameters: $a = 0.0$, $b = 0.0$, $c = 0.0$, $d = -1.0$.

Equation:

$$x_1' = a + bx_1 + cx_1^2 + dx_1^3 .$$

From a qualitative point of view, the dynamics of a one-dimensional differential equation are quite simple: either an orbit eventually approaches an equilibrium point, or it goes to infinity. It is important, however, to understand the dependence of the dynamics of an equation on its parameters because the number of equilibria and their stability types may change as these parameters are varied. Despite the apparent simplicity of this example, the three most common qualitative changes around an equilibrium point -- the so-called pitchfork, saddle-node and transcritical bifurcations (see section 7.2) -- can be seen in *cubic1d*.

References: Arnold [1983] p. 279; Guckenheimer & Holmes [1983] p. 145.

7.2. 2D Differential Equations

In the exploration of two-dimensional differential equations, it is very helpful to use the *DirField* (direction field) and the *Flow* entries on the UTILITIES menu (see section 6.2).

---

linear2d *General two-dimensional linear system.*

Type: 2D Differential.

Parameters: $a = 0.0$, $b = 1.0$, $c = -1.0$, $d = 0.0$.

Equations:

$$x_1' = ax_1 + bx_2$$

$$x_2' = cx_1 + dx_2 .$$

The study of linear two-dimensional systems forms the beginning of the geometric theory of differential equations. The chief importance of linear systems stems from the fact that the phase portrait of a nonlinear system around an equilibrium point looks qualitatively like that of its linearization (except possibly in the presence of purely imaginary eigenvalues).

From an analytical point of view, it is quite easy to solve these equations using eigenvalue and eigenvector methods. You should investigate by varying the parameters in this system all possible qualitatively different phase portraits around the unique equilibrium (0, 0), as classified in the references.

As we have discussed in Chapter 2, solving differential equations numerically can be tricky. This is true even in the case of linear systems, if the equations are "stiff". For example, try to solve the system with the parameter values $a = 998.0$, $b = 1998.0$, $c = -999.0$, $d = -1999.0$ using *Euler* and *Runge-Kutta*, and various step sizes (see Forsythe et al.).

References: Arnold [1973] p. 133; Braun [1983] p. 416; Boyce & DiPrima [1977] p. 397; Forsythe et al. [1977] p. 124; Hirsch & Smale [1974] p. 89.

---

pendulum *Nonlinear pendulum on the plane.*

Type: 2D Differential.

Parameters: $c = 0.0$, $g = 1.0$, $l = 1.0$, $m = 1.0$.

Equations:

$$x_1' = x_2$$

$$x_2' = -\frac{g}{l}\sin(x_1) - \frac{c}{lm}x_2 .$$

This is probably the best-known example from classical mechanics. When there is no damping ($c = 0.0$), the rest position at the origin is a stable equilibrium of *center* type; that is, the phase portrait near the origin looks like that of the linear harmonic oscillator. However, in contrast to the linear harmonic oscillator, the non-linear term in the *pendulum* equation gives rise to periodic orbits that do not have constant phase. You should run two solutions simultaneously, starting near the origin, and observe that their corresponding periods are different. The vertical position of the pendulum $(\pi, 0)$ is also an equilibrium, but of saddle type; see *Figure 7.1.*

If we add damping ($c > 0.0$), then the origin becomes a stable spiral, and the pendulum eventually stops; see *Figure 7.2.* You should investigate the effects of mass (m), length (l), and the gravitational constant (g) on the qualitative motion of the pendulum.

In most books, including this one, the phase space of the planar pendulum is taken to be $\mathbf{R}^2$. From the mathematical point of view, it is sometimes advantageous to study these equations on the cylinder $S^1 \times \mathbf{R}$ because they are invariant under the transformation $x_1 \rightarrow x_1 + 2\pi$. See Arnold for the phase portrait of the pendulum on the cylinder.

References: Arnold [1980] p. 90; Boyce & DiPrima [1977] p. 413; Hirsch & Smale [1974] p. 183.

---

vanderpol *Oscillator of van der Pol: A unique limit cycle.*

Type: 2D Differential.

Parameters: $a = 1.0$.

Equations:

$$x_1' = x_2 - (x_1^3 - ax_1)$$

$$x_2' = -x_1 .$$

These equations are the Lienard form of the van der Pol's differential equations modeling, an electrical circuit whose resistive properties change as a non-linear function of the current. (For the original van der Pol's version, see the equation *forcevdp*.) All solutions (except the origin) of these equations eventually converge to the same unique periodic orbit and oscillate forever; see *Figure 7.3*. Such an isolated periodic orbit is called a (stable) *limit cycle*.

Furthermore, orbits near the stable limit cycle have the same asymptotic period as the period of the limit cycle. They are also asymptotically "in phase" with the limit cycle. You should run two solutions of the equation *vanderpol* simultaneously (turn the flasher on), and compare your results to those for the *predprey* equations.

As the parameter a is increased through zero, the equation *vanderpol* undergoes an important qualitative change, called a *Hopf bifurcation*. When $a < 0$ the origin is globally attracting at an exponential rate. For $a = 0$ the origin is still attracting, but not at an exponential rate. Finally, when a becomes positive, a unique limit cycle is born, which grows larger with increasing a. See the equation *hopf* for more information on this important bifurcation.

References: Boyce & DiPrima [1977] p. 448; Hale [1969] p. 60; Hirsch & Smale [1974] pp. 217, 278.

---

predprey *Predator-prey equations, competing species, etc.*

Type: 2D Differential.

Parameters: $a = 1.0$, $b = 1.0$, $c = 1.0$, $d = 1.0$, $m = 0.0$, $n = 0.0$.

Equations:

$$x_1' = (a - bx_2 - mx_1)x_1$$

$$x_2' = (cx_1 - d - nx_2)x_2 .$$

This system of equations, describing the evolution of two interacting species, is one of the pioneering examples in mathematical ecology. When $m = n = 0.0$ and all the remaining parameters are positive, the equations are called the predator-prey model of Volterra and Lotka. The qualitative behavior of the system is oscillatory: every trajectory in the positive quadrant is a periodic orbit encircling the unique equilibrium; see *Figure 7.4*. The phase portrait looks much like that of a linear harmonic oscillator around a *center* type equilibrium. However, the nonlinearity of the predator-prey equations gives rise to periodic orbits that do not have constant phase. You should run two initial conditions simultaneously, and observe that their corresponding orbits have different periods (see also the equation *pendulum*).

When all the parameters, including m and n, are positive, the equations are used to model two species with limited growth (also called competing species). Every trajectory eventually approaches either an equilibrium point or a limit cycle. Hence, there are absolute upper bounds that neither species can exceed, regardless of the initial conditions.

References: Boyce & DiPrima [1973] p. 421; Hirsch & Smale [1974] p. 255; Maynard Smith [1968] p. 46.

---

saddlenod *Saddle-node: Generic bifurcation of an equilibrium point.*

Type: 2D Differential.

Parameter: $a = 0.0$.

Equations:

$$x_1' = a - x_1^2$$

$$x_2' = -x_2 .$$

Local bifurcations around an equilibrium point occur when at least one eigenvalue of the linearized system has zero real part. The next four examples are typical bifurcations occurring when equilibria of differential equations that depend on a single parameter become structurally unstable in a "mild" way.

For the parameter values $a > 0$, the equation *saddlenod* has two equilibria near the origin, one of which is a saddle and the other a node. The parameter value $a = 0$ is the bifurcation value where the two equilibria merge into one semistable equilibrium called saddle-node. Finally, for $a < 0$, the equilibrium at the origin vanishes; see *Figures 7.5-7.* This qualitative change in the phase portrait near the origin is called a *saddle-node bifurcation.* Saddle-node is the *generic* bifurcation of an equilibrium point at which there is a simple zero eigenvalue. Here, generic means that in "almost all" one-parameter differential equations, an equilibrium point satisfying the eigenvalue condition above undergoes a saddle-node bifurcation.

The "center manifold theorem" (cf. Carr) allows one to reduce the study of the saddle-node bifurcation to a one-dimensional differential equation. In fact, in this two-dimensional example, the equations are uncoupled and the saddle-node bifurcation takes place in the x_1 variable (see the equation ***cubic1d***). The second equation is added for visual effects only.

These equations are the simplest example exhibiting saddle-node bifurcation. You should add some more non-linear terms to the equations and study their effects on the dynamics.

References: Carr [1980]; Chow & Hale [1982] pp. 16, 327, 364; Guckenheimer & Holmes [1983] pp. 124, 146.

---

transcrit

Transcritical bifurcation: Exchange of stability.

Type: 2D Differential.

Parameter: $a = 0.0$.

Equations:

$$x_1' = ax_1 - x_1^2$$

$$x_2' = -x_2 .$$

As in the previous example, this one-parameter differential equation has, for the parameter value $a = 0$, an equilibrium at the origin with a simple zero eigenvalue. If we assume further that the origin remains an equilibrium point for all small values of the parameter, then generically the equilibrium point undergoes a *transcritical bifurcation.* When $a < 0$, the origin is stable, but there is also an unstable equilibrium. At the parameter value $a = 0$, the two equilibria merge at the origin, which becomes semistable. For $a > 0$, the origin becomes unstable and the other equilibrium reappears as a stable equilibrium.

Reference: Guckenheimer & Holmes [1983] pp. 145, 149.

---

pitchfork

Bifurcation of an equilibrium with reflection symmetry.

Type: 2D Differential.

Parameter: $a = 0.0$.

Equation:

$$x_1' = ax_1 - x_1^3$$

$$x_2' = -x_2 .$$

At an equilibrium point with a simple zero eigenvalue, generically one expects to encounter the saddle-node bifurcation. If this is not the case in a specific system, then there is probably something special about the equations which prevents the occurrence of the saddle-node bifurcation. For instance, in the previous example, the equation *transcrit*, the origin was assumed to be an equilibrium for all values of the parameter a. Another special case is the presence of symmetry. If we assume, in addition to the eigenvalue condition above, that the right-hand side of differential equations have reflection symmetry, $\mathbf{f}(-\mathbf{x}) = -\mathbf{f}(\mathbf{x})$, then generically the equilibrium point undergoes a *pitchfork bifurcation.* In this example of pitchfork bifurcation, when $a < 0$, the origin is a stable equilibrium. At $a = 0$ the origin loses it stability. Finally, for $a > 0$, two new stable equilibrium points in addition to the origin appear.

For an example of the pitchfork bifurcation in three dimensions, you should study the *lorenz* equations near the parameter value $r = 1.0$.

Reference: Guckenheimer & Holmes [1983] pp. 147, 149.

---

hopf

Hopf bifurcation: Birth of a periodic orbit.

Type: 2D Differential.

Parameter: $a = 1.0$.

Equations:

$$x_1' = -x_2 + x_1(a - (x_1^2 + x_2^2))$$

$$x_2' = x_1 + x_2(a - (x_1^2 + x_2^2)) .$$

Consider a system of differential equations depending on a parameter a that has an equilibrium point with a simple pair of purely imaginary eigenvalues for the parameter value $a = 0.0$. The implicit function theorem guarantees that there will be a unique equilibrium point for all small values of a near zero. So, as the parameter a is varied, the number of equilibria does not change. (Compare this behavior with, for example, the saddle-node bifurcation.) However, if the eigenvalues cross the imaginary axis for $a = 0$, then the dimensions of the stable and the unstable manifolds of the equilibrium change. (The *stable manifold* of an equilibrium is the set of points $\mathbf{x}$ in the phase space such that trajectories through $\mathbf{x}$ tend towards the equilibrium; the *unstable manifold* is the set of points $\mathbf{x}$ such that the trajectories through $\mathbf{x}$ tend towards the equilibrium in reverse time.) Generically, this qualitative change is marked by the birth of a periodic orbit near the equilibrium point. It is quite easy to detect the Hopf bifurcation by analyzing the dependence of the eigenvalues of the linearized system on the parameter. However, determining the stability type of the periodic orbit may be very tedious.

In this example, when $a < 0$, the origin is a stable equilibrium point, attracting nearby orbits at an exponential rate. At the value $a = 0$, the origin is still stable, but the attraction rate is no longer exponential. For $a > 0$, the origin becomes unstable and an attracting periodic orbit surrounding the origin is born; see *Figures 7.8-10*. The radius of the periodic orbit grows with increasing a. These equations are the simplest example exhibiting Hopf bifurcation; in fact, they can be solved exactly in polar coordinates. You should investigate the effects of nonlinear perturbations of this system.

The Hopf bifurcation is very important in applications; see, for example, the equation *vanderpol*. In addition, a more complicated Hopf bifurcation than the one described above takes place in the equation *lorenz* for the parameter value $r = 24.74$. In this variation of the Hopf bifurcation, called *subcritical Hopf bifurcation*, an equilibrium point loses its stability by absorbing a non-stable periodic orbit.

References: Braun [1983] p. 430; Chow & Hale [1982] pp. 9, 338; Guckenheimer & Holmes [1983] pp. 146, 150; Hassard et al. [1981]; Marsden & McCracken [1976].

---

dzero1

Generic unfolding of a double zero eigenvalue.

Type: 2D Differential.

Parameters: $a = -0.2$, $b = 0.2$, $c = 1.0$, $d = 1.0$.

Equations:

$$x_1' = x_2$$

$$x_2' = a + bx_2 + cx_1^2 + dx_1x_2 \,.$$

These equations are a *versal deformation* (also called *unfolding*) around a degenerate equilibrium of a vector field whose linear part is $\begin{bmatrix} 0 & 1 \\ 0 & 0 \end{bmatrix}$. "In other words, a generic two-parameter family of differential equations in the plane having a singular point with two vanishing roots of the characteristic equation for some value of the parameter can be reduced to the form indicated above by a continuous change of the parameters and a continuous change of the phase coordinates continuously depending on the parameters" (from Arnold).

With suitable scaling and coordinate transformations, two of the parameters, c and d, can be reduced to the cases $c = 1$ and $d = \pm 1$, if they are nonzero. Therefore, the system is a bifurcation of codimension two (depending on two parameters), with a and b as the versal deformation parameters which vary in a small neighborhood of zero. For further information on this system, consult the references below.

References: Arnold [1983] p. 280; Guckenheimer & Holmes [1983] p. 364.

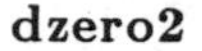

dzero2

Unfolding of double zero eigenvalue with origin fixed.

Type: 2D Differential.

Parameters: $a = -0.2$, $b = 0.2$, $c = 1.0$, $d = 1.0$.

Equations:

$$x_1' = x_2$$

$$x_2' = ax_1 + bx_2 + cx_1^2 + dx_1x_2 .$$

These equations are a versal deformation around a degenerate equilibrium of a vector field whose linear part is $\begin{bmatrix} 0 & 1 \\ 0 & 0 \end{bmatrix}$. This versal deformation is much like the previous example, *dzero1*, except that the origin is required to remain an equilibrium point under all perturbations.

With suitable scaling and coordinate transformations, two of the parameters, c and d, can be reduced to the cases $c = 1$ and $d = \pm 1$, if they are nonzero. Therefore, the system is a bifurcation of codimension two, with a and b as the versal deformation parameters which vary in a small neighborhood of zero. For further details, consult the reference below.

Reference: Chow & Hale [1982] p. 444.

---

dzero3

Unfolding of double zero eigenvalue with symmetry.

Type: 2D Differential.

Parameters: $a = -0.2$, $b = 0.2$, $c = 1.0$, $d = -1.0$.

Equations:

$$x_1' = x_2$$

$$x_2' = ax_1 + bx_2 + cx_1^3 + dx_1^2x_2 .$$

These equations are a versal deformation around a degenerate equilibrium of a *rotationally symmetric* vector field whose linear part is $\begin{bmatrix} 0 & 1 \\ 0 & 0 \end{bmatrix}$. With suitable scaling and coordinate transformations, two of the parameters, c and d, can be reduced to the cases $c = \pm 1$ and $d = -1$, if they are nonzero. Therefore, the system is a bifurcation of codimension two, with a and b as the versal deformation parameters which vary in a small neighborhood of zero. As the deformation parameters are varied, these equations go through a very complicated sequence of bifurcations. For further details, consult the references below.

References: Carr [1981] p. 57; Guckenheimer & Holmes [1983] p. 371; Takens [1974].

---

hilbert2

A quadratic system with two limit cycles.

Type: 2D Differential.

Parameter: $a = 0.8$.

Equations:

$$x_1' = P\ cos(a) - Q\ sin(a)$$

$$x_2' = P\ sin(a) + Q\ cos(a),$$

where

$$P = 169(x_1 - 1)^2 - 16(x_2 - 1)^2 - 153$$

$$Q = 144(x_1 - 1)^2 - 9(x_2 - 1)^2 - 135$$

In 1900, as part of problem sixteen of his famous list, Hilbert posed the following question: what is the number of limit cycles (isolated periodic orbits) of a general polynomial system of differential equations on the plane? Although the problem remains unsolved, it has nevertheless enjoyed a colorful history. The first major result was the "theorem" of Dulac asserting the finiteness of the total number of limit cycles, without a specific bound. Recently, however, a gap in his "proof" was discovered, one which has turned out to be nonrectifiable. Consequently, for a general polynomial system, even the question of finiteness continues unanswered.

The quadratic systems, as the first nontrivial case of Hilbert's problem, have been investigated intensively. In the fifties, Petrovskii and Landis claimed that the maximum number of limit cycles for quadratic systems was three. Unfortunately, their long and difficult papers --

however influential -- were plagued with serious errors. In 1980, both Shi Song-ling and, independently, Wang Mingshu resolved the mystery by producing a quadratic example with at least four limit cycles (see the equation *hilbert4*). Most recently, the theorem of Dulac has been partially salvaged by Bamon: in the case of quadratic systems, the number of limit cycles is indeed finite.

Yeh has shown that the quadratic system above has two limit cycles: an attracting one around (0, 0) and a repeller around (2, 2), as seen in *Figure 7.11*. The repelling limit cycle can be located by following a nearby orbit while using a negative step size.

References: Bamon [1984]; Chicone & Tian [1982]; Yeh [1958], [1982], [1985].

---

hilbert4

A quadratic system with four limit cycles.

Type: 2D Differential.

Parameters: $a = 0.0$, $b = 0.0$, $c = 0.0$.

Equations:

$$x_1' = ax_1 - x_2 - 10.0x_1^2 + (5.0 + b)x_1x_2 + x_2^2$$

$$x_2' = x_1 + x_1^2 + (8.0c - 25.0 - 9.0b)x_1x_2 .$$

It was proved by Shi Song-ling that for the parameter values $a = -10^{-200}$, $b = -10^{-13}$, and $c = -10^{-52}$, the quadratic system above has at least four limit cycles. It is difficult to draw pictures of these limit cycles using numerical solutions because of large variations in the magnitude of the coefficients. The parameter values above have been chosen to make the proof work. In fact, there could be a "fat" subset (bigger than a point!) of the parameter space for which four limit cycles exist. By searching in the parameter space, it may be possible to find such a subset that might also include coefficients more amenable to numerical computations.

References: Chicone & Tian [1982]; Shi [1980]; Yeh [1982], [1985].

---

averfvdp

Averaged forced van der Pol oscillator.

Type: 2D Differential.

Parameters: $g = 0.3$, $s = 0.15$.

Equations:

$$x_1' = x_1 - sx_2 - x_1(x_1^2 + x_2^2)$$

$$x_2' = sx_1 + x_2 - x_2(x_1^2 + x_2^2) - g \; .$$

One of the important techniques for determining periodic and almost periodic solutions of differential equations is the so-called method of *averaging* (cf. Hale [1969]). It is particularly useful in the study of periodically forced two-dimensional systems because the averaged equations give approximations to Poincare maps. Moreover, the averaged equations are two-dimensional autonomous systems, which, at least in theory, are possible to analyze almost completely.

The system above is the averaged equations of the periodically forced oscillator of van der Pol (see the equations *vanderpol* and *forcevdp*). Hyperbolic equilibria and periodic orbits of the averaged equations correspond to periodic orbits and invariant tori, respectively, of the periodically forced van der Pol's oscillator, *forcevdp*; see *Figure 7.12.*

References: Guckenheimer & Holmes [1983] p. 70; Hale [1969] p. 186; Holmes & Rand [1978].

---

gradient

Universal unfolding of the elliptic umbilic.

Type: 2D Differential.

Parameters: $a = 0.0$, $b = 0.0$, $c = 0.0$.

Equations:

$$x_1' = x_1 x_2 - a - 2.0cx_1$$

$$x_2' = 0.5(x_1^2 - x_2^2) - b - 2.0cx_2 \; .$$

There are subtle relationships between Thom's catastrophe theory of the unfolding of a potential function, and the unfolding of the corresponding vector field. The one-dimensional case is straightforward because the dynamics of a one-dimensional differential equation are determined by its equilibrium points and their stability types, and singularity theory provides the tools for a systematic study of the zeros of functions. In two or more dimensions, however, the relationship is not straightforward, as illustrated by the elliptic umbilic.

The differential equations above are the *gradient vector field* (see Hirsh & Smale for the definition of gradient vector field, and an elementary example) generated by the potential function

$$V = \frac{1}{2}\left(\frac{x_2^3}{3} - x_1^2 x_2\right) + a x_1 + b x_2 + c\,(x_1^2 + x_2^2)\,,$$

which is the universal unfolding (the most general perturbation) of the elliptic umbilic. These differential equations possess *heteroclinic orbits* (a heteroclinic orbit is an orbit which tends to one equilibrium point in forward time and to another equilibrium point in reverse time), which undergo bifurcations that are not detectable from the potential function.

References: Chow & Hale [1982] p. 450; Guckenheimer & Holmes [1983] p. 357; Hirsh & Smale [1974] p. 199.

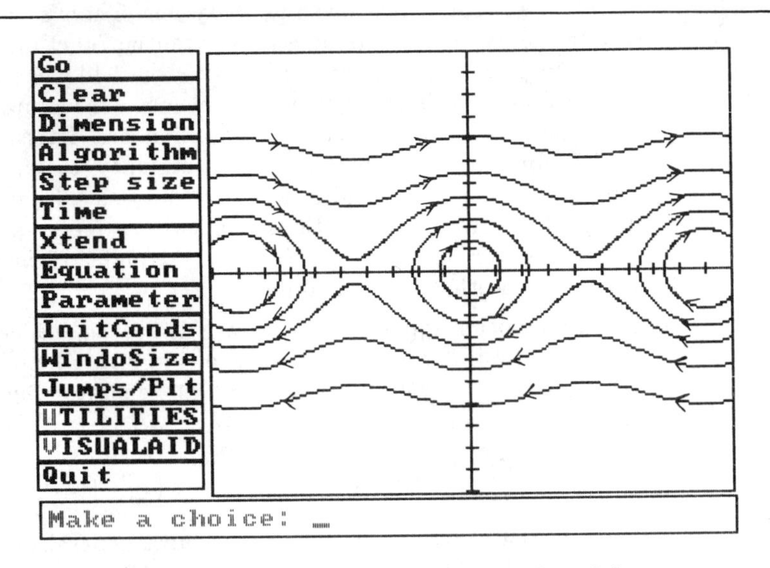

Figure 7.1. Phase plane of the undamped pendulum.

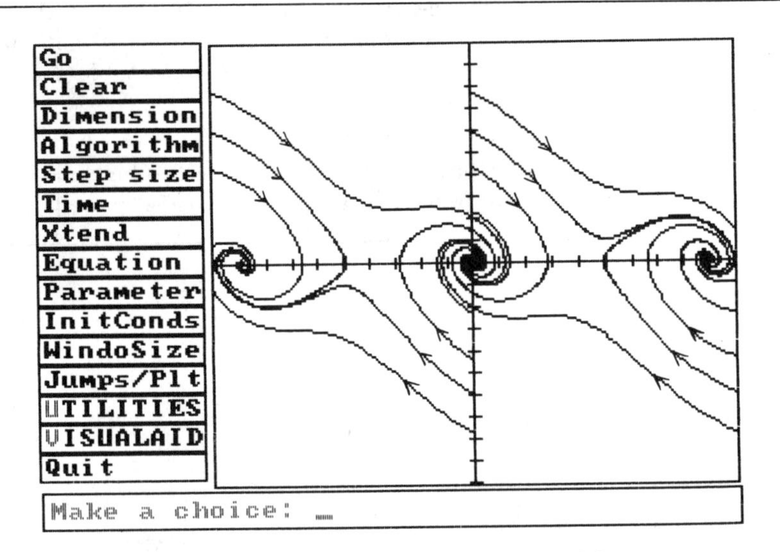

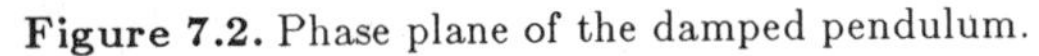
Figure 7.2. Phase plane of the damped pendulum.

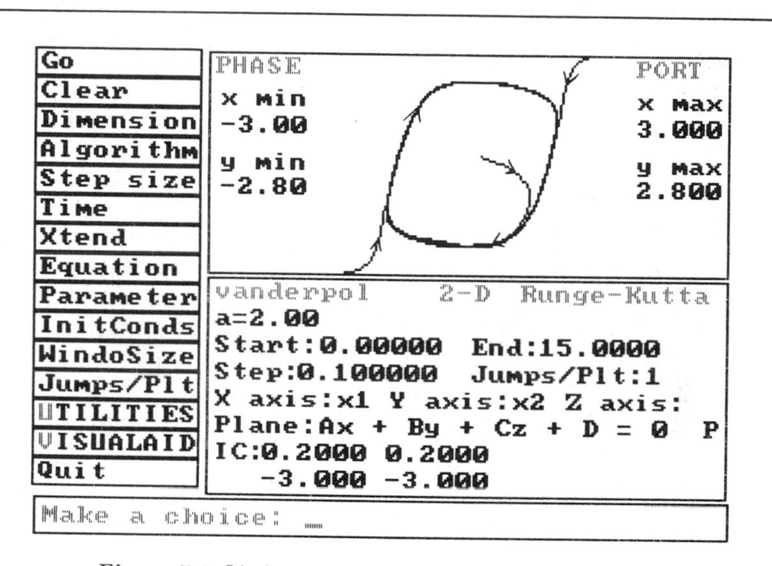

Figure 7.3. Limit cycle of the oscillator of van der Pol.

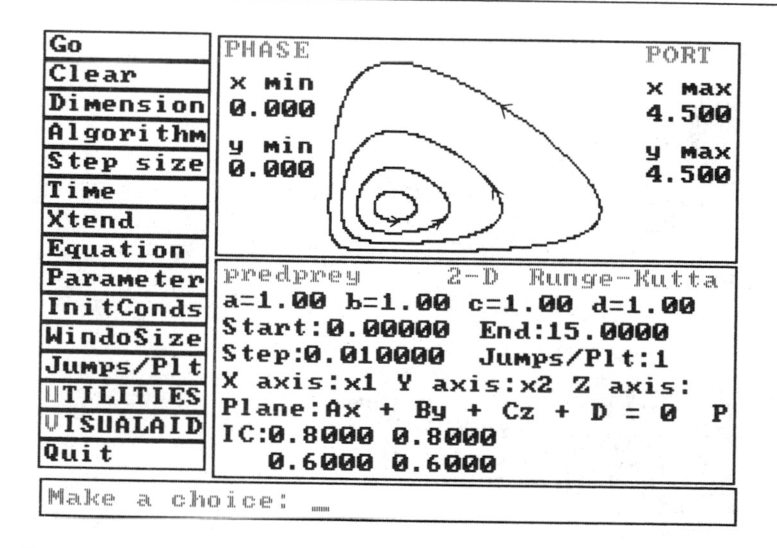

Figure 7.4. Phase plane of the *predprey* equations of Volterra and Lotka.

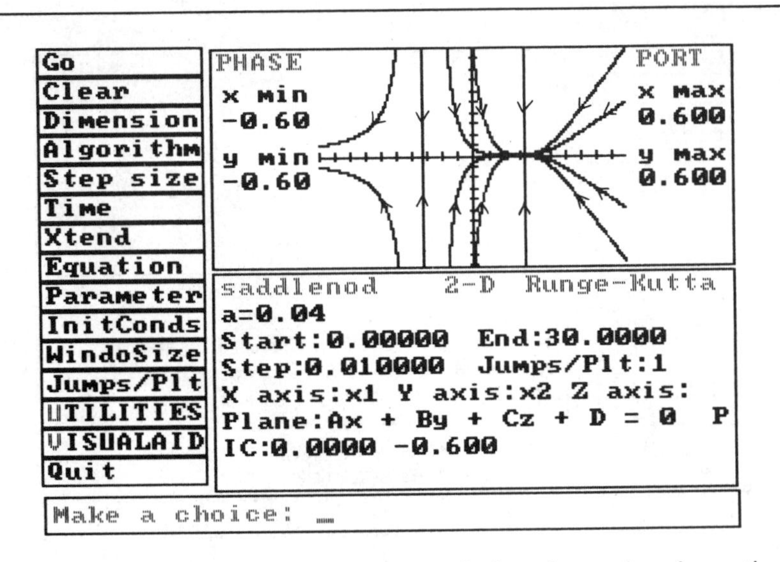

Figure 7.5 Two equilibria, one saddle and the other node, of equation *saddlenod*. The next two figures illustrate the saddle-node bifurcation as the parameter a is varied.

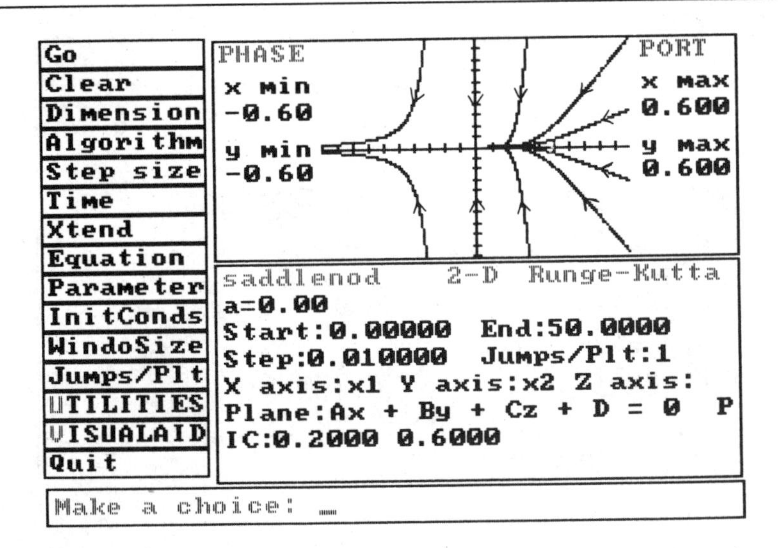

Figure 7.6. The two equilibria of Figure 7.5 merge at the origin, becoming a single semistable equilibrium.

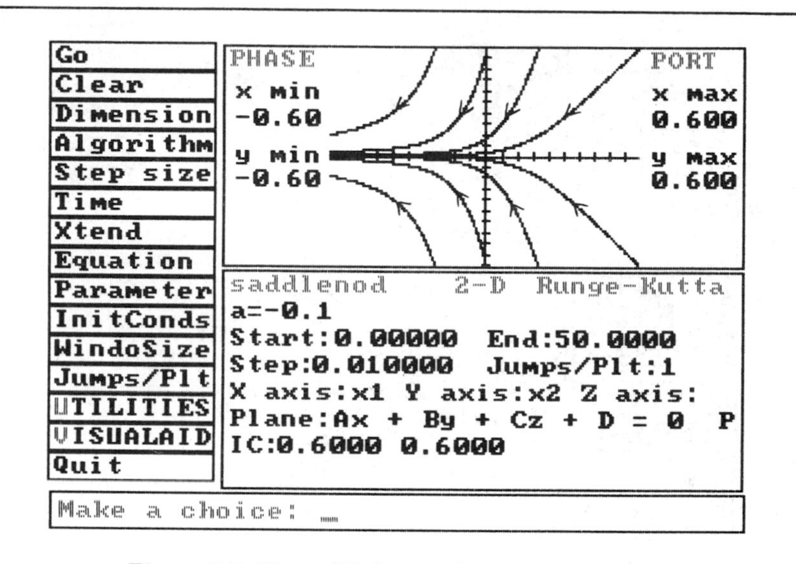

Figure 7.7. The equilibrium at the origin disappears.

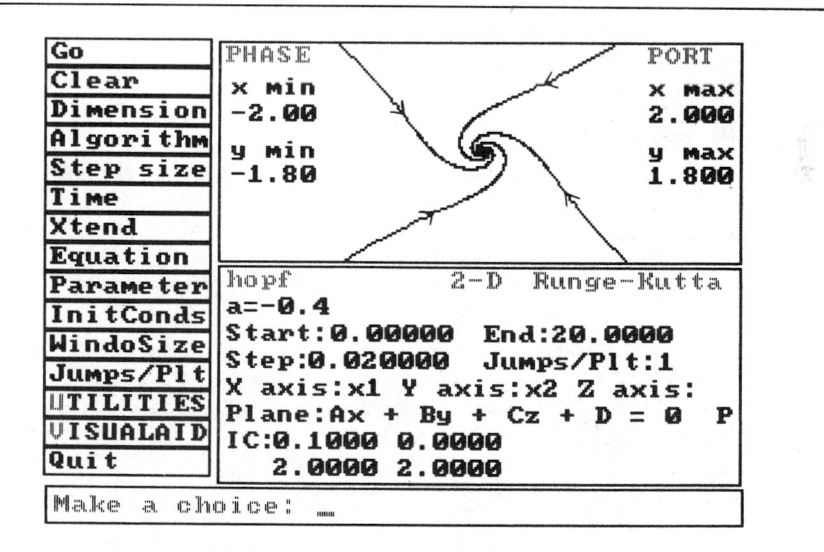

Figure 7.8 Stable equilibrium of the equation *hopf*. The next two figures illustrate the Hopf bifurcation.

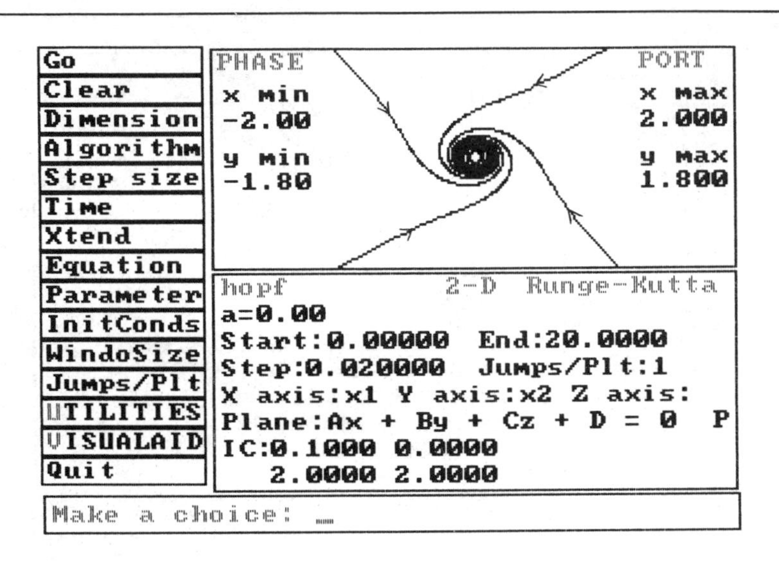

Figure 7.9. The origin is still stable, but no longer attracting at an exponential rate.

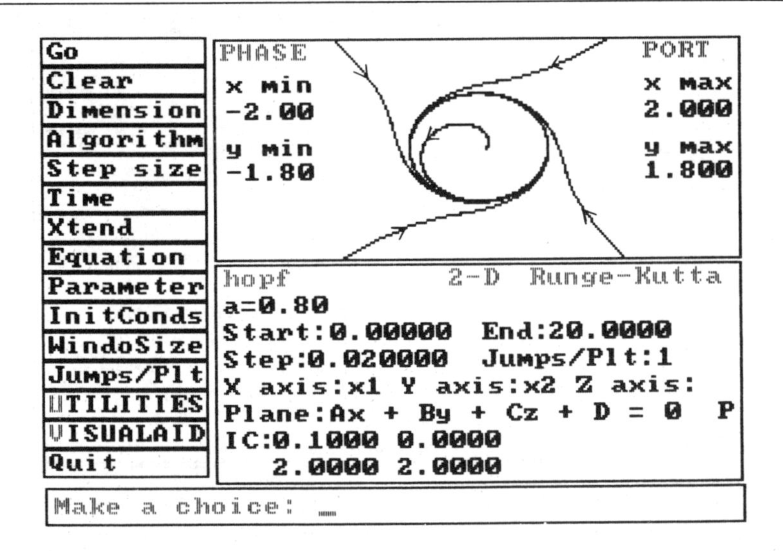

Figure 7.10. The origin becomes unstable and a stable periodic orbit is born.

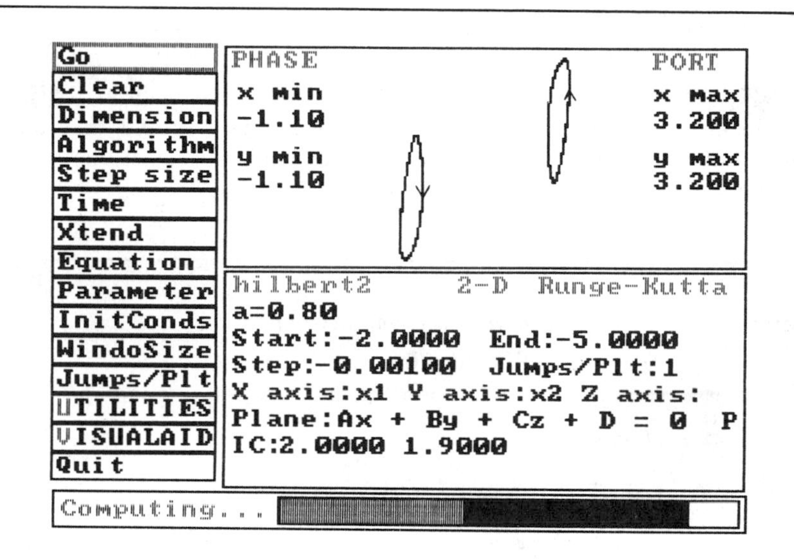

Figure 7.11. The two limit cycles of the planar quadratic system *hilbert2*.

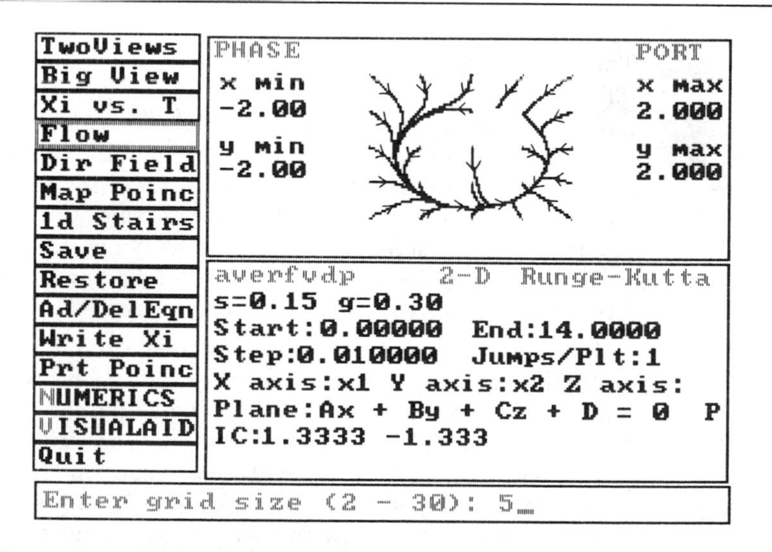

Figure 7.12. Phase portrait of the averaged forced van der Pol's equation, *averfvdp*, plotted using the *flow* entry on the *UTILITIES* menu.

7.3. 3D Differential Equations

PHASER provides a large collection of graphical utilities for studying solutions of equations in three dimensions. You should take advantage of not only the standard rotations, perspective projection, etc., but also of the *CutPlane* and *MapPoinc* views.

---

lorenz

The most famous strange attractor.

Type: 3D Differential.

Parameters: $b = 2.66666666$, $r = 28.0$, $s = 10.0$.

Equations:

$$x_1' = s\,(-x_1 + x_2)$$

$$x_2' = rx_1 - x_2 - x_1x_3$$

$$x_3' = -bx_3 + x_1x_2 \,.$$

The Lorenz equations are undoubtedly the most famous differential equations of the past decade, with major impact on both theoretical and applied dynamical systems. I therefore made the default equation in the 3D library *lorenz* so that everyone can see them at least once. Another good reason for this choice is that the strange attractor (an attracting set that is more complicated than an equilibrium point or a limit cycle) of Lorenz could not have been discovered without the computer.

This system of equations is a three-mode approximation to the motion of a layer of fluid heated from below. The parameter r corresponds to the Reynolds number, and as this parameter is varied, the system goes through a remarkable sequence of qualitative changes. For the parameter values indicated above, almost all solutions converge to a strange attractor; moreover, once on the attractor they exhibit almost random behavior; see *Figure 7.13.* Furthermore, two solutions starting close to each other do rather different things in a short period of time; see *Figure 7.14.* The sensitive dependence on initial conditions has important consequences for applications: in fact, it appears to be the motivation for some of the most profound investigations of Lorenz, a mathematician turned meteorologist. You should try this experiment using PHASER, but be forewarned that most of the observed properties of this system still await a proof, despite the determined efforts of many mathematicians.

References: Guckenheimer & Holmes [1983] p. 92; Lorenz [1963]; Sparrow [1982].

---

linear3d

General three-dimensional linear system.

Type: 3D Differential.

Parameters: $a = 0.0$, $b = -1.0$, $c = 0.0$, $d = 1.0$, $e = 0.0$, $f = 0.0$, $g = 0.0$, $h = 0.0$, $i = -0.2$.

Equations:

$$x_1' = ax_1 + bx_2 + cx_3$$

$$x_2' = dx_1 + ex_2 + fx_3$$

$$x_3' = gx_1 + hx_2 + ix_3 .$$

The remarks for *linear2d* apply to this system as well. However, visualizing an orbit in three-space on a two-dimensional screen may require experimentation. Therefore, you should use these equations to improve your understanding of the 3D graphical facilities of PHASER.

Reference: Arnold [1973] p. 139.

---

vibration

Periodically forced linear vibrations.

Type: 3D Differential.

Parameters: $a = 1.0$, $c = 0.0$, $f = 0.0$, $k = 1.0$, $m = 1.0$.

Equations:

$$x_1' = x_2$$

$$x_2' = -\frac{c}{m}x_2 - \frac{k}{m}x_1 + \frac{f}{m}\cos(ax_3)$$

$$x_3' = 1.0 .$$

This system of equations, which has important engineering applications in mechanics, electrical circuits, etc., describes the motion of a periodically forced linear oscillator (see section 1.4).

Suppose first that there is no damping ($c = 0.0$) and no forcing ($f = 0.0$). This, then, is just the linear harmonic oscillator whose solutions are all periodic. If we apply a periodic force to the system ($f > 0.0$), then solutions in general look like wave packets, reflecting the natural frequency ($a/2\pi$) of the unforced system and the forcing frequency ($\sqrt{k/m}/2\pi$); see *Figure 7.15*. The most interesting case, when the two frequencies are equal, is known as the *resonance* case. In resonance the system produces unbounded oscillations; see *Figure 7.16*.

If there is any damping in the system ($c > 0.0$), then the qualitative behavior is simple: all solutions eventually go to zero.

References: Section 1.4; Boyce & DiPrima [1977] p. 135; Braun [1983] p. 165; Guckenheimer & Holmes [1983] p. 27.

---

bessel

Bessel's equation.

Type: 3D Differential.

Parameter: $n = 0.0$.

Equations:

$$x_1' = x_2$$

$$x_2' = -\frac{x_2}{x_3} - (1.0 - \frac{n^2}{x_3^2})x_1$$

$$x_3' = 1.0 .$$

Both Bessel's equation and the next four equations are examples from the theory of "special functions" of mathematical physics. They are second-order nonautonomous equations, but we have converted them to equivalent first-order systems for computational purposes.

Because of their theoretical and practical importance, these equations were studied intensively during the last century by a great number of researchers, including prominent mathematicians.

Though there are much better ways than using numerical methods to compute some of the solutions of these equations, it is still instructive to see their pictures and to test the accuracy of numerical algorithms on explicitly known solutions, especially around singular points.

Solutions of Bessel's equation are known as Bessel functions; see *Figure 7.17.* They have many diverse applications to problems in physics and engineering -- e.g., wave propagation, elasticity, fluid mechanics -- and especially in problems having cylindrical symmetry.

References: Boyce & DiPrima [1977] p. 195; Simmons [1972] p. 232.

---

euler

Euler's equation.

Type: 3D Differential.

Parameters: a =1.0, b = 1.0.

Equations:

$$x_1' = x_2$$

$$x_2' = -\frac{1.0}{x_3}(ax_2 - \frac{b}{x_3}x_1)$$

$$x_3' = 1.0 \,.$$

Euler's equation, or the equidimensional equation, is one of the simplest examples of a differential equation with a regular singular point. You should first study the analytical treatment given in the references, and then compare the results of numerical simulations with the explicit solutions.

References: Boyce & DiPrima [1977] p. 174; Braun [1983] p. 196.

---

laguerre

Laguerre's equation.

Type: 3D Differential.

Parameter: $p = 3.0$.

Equations:

$$x_1' = x_2$$

$$x_2' = -\frac{1.0}{x_3}((1.0 - x_3)x_2 + px_1)$$

$$x_3' = 1.0 .$$

Laguerre's equation is important in the study of the quantum mechanics of the hydrogen atom. When the parameter p is a nonnegative integer m, there is a polynomial solution $L_m(t)$ called Laguerre's polynomial of order m; see *Figure 7.18*. The first four Laguerre polynomials are:

$$L_0(t) = 1 ,$$

$$L_1(t) = 1 - t ,$$

$$L_2(t) = 2 - 4t + t^2 ,$$

$$L_3(t) = 6 - 18t + 9t^2 - t^3$$

You should plot them using PHASER.

Reference: Boyce & DiPrima [1977] p. 185.

---

legendre

Legendre's equation.

Type: 3D Differential.

Parameter: $p = 2.0$.

Equations:

$$x_1' = x_2$$

$$x_2' = \frac{1.0}{1.0 - x_3^2}(2.0x_2x_3 + p(p + 1.0)x_1)$$

$$x_3' = 1.0 .$$

Legendre's equation appears in the study of the potential equation in spherical coordinates. When the parameter p is a nonnegative integer m, there is a polynomial solution $P_m(t)$, with $P_m(1) = 1$, which is called the Legendre polynomial of order m. These polynomials were used by Legendre in his investigations of the gravitational attraction of ellipsoids. The first four Legendre polynomials are:

$$P_0(t) = 1 ,$$

$$P_1(t) = t ,$$

$$P_2(t) = \frac{3}{2}t^2 - \frac{1}{2} ,$$

$$P_3(t) = \frac{5}{2}t^3 - \frac{3}{2}t .$$

You should plot them using PHASER.

References: Boyce & DiPrima [1977] pp. 101, 167; Simmons [1972] p. 219.

---

forcevdp *Periodically forced van der Pol's oscillator.*

Type: 3D Differential.

Parameters: $a = 0.3$, $b = 0.0$, $c = 1.0$.

Equations:

$$x_1' = x_2$$

$$x_2' = -x_1 - a(x_1^2 - 1.0)x_2 + b\cos(cx_3)$$

$$x_3' = 1.0 \qquad (mod\ \frac{2\pi}{c}) .$$

This is the oscillator of van der Pol with a periodic forcing term. It is a nonautonomous system in $\mathbf{R}^2$. However, the appropriate phase space to use is $\mathbf{R}^2 \times S^1$ since the third variable, time, figures into the system only as the argument of a periodic function.

You should first study the unforced equation ($b = 0.0$) which, through a change of variables, is equivalent to the equation *vanderpol*. In this case, the system has a globally attracting limit cycle on the (x_1, x_2)-plane. Therefore, when the system is forced, the solutions are attracted to an invariant torus ($S^1 \times S^1$). As the forcing amplitude and the frequency are varied, the flow on the torus undergoes a complex series of bifurcations. Eventually the torus starts folding onto itself, giving rise to a strange attractor. Many of the local bifurcations can be explained in terms of the averaged equation, as discussed in equation *averfvdp*. In numerical simulations, it is instructive to study the globally defined planar Poincare map $x_3 = 0.0$.

References: Guckenheimer & Holmes [1983] p. 67; Hale [1969] p. 194.

---

forcepen

Periodically forced pendulum equation.

Type: 3D Differential.

Parameters: $a = 0.5$, $b = 0.1$.

Equations:

$$x_1' = x_2$$

$$x_2' = -(a^2 + b\cos(x_3))\sin(x_1)$$

$$x_3' = 1.0 \qquad (mod\ 2\pi)\ .$$

These equations describe the motion of a pendulum (see the equation *pendulum*) with a periodically excited supporting pivot. Notice that for all values of the parameter b there are periodic orbits given by $(0, 0; x_3(t))$ and $(\pi, 0; x_3(t))$, corresponding to the equilibrium positions of the unforced pendulum. Linearizing the vector field about these periodic orbits yields the equation of Mathieu (see the equation *mathieu* below). Using *Floquet theory* (cf. Hale [1969]), it can be shown that the first periodic orbit, corresponding to the lowest equilibrium of the pendulum, is stable when b is sufficiently small and $a \neq n/2$, where n is any integer. When $b > 0$ and $a = 1/2$, the period of this periodic orbit doubles (second harmonic). This phenomenon is called *parametric resonance*. The second periodic orbit is of saddle type for b non-zero and sufficiently small, and all $a \neq 0$.

References: Guckenheimer & Holmes [1983] pp. 29, 204; Hale [1969] p. 118.

---

mathieu

Mathieu's equation.

Type: 3D Differential.

Parameters: $a = 0.5$, $b = 0.1$.

Equations:

$$x_1' = x_2$$

$$x_2' = -(a^2 + b\cos(x_3))\,x_1$$

$$x_3' = 1.0 \qquad (mod\ 2\pi).$$

This is the linearization of the forced pendulum equations about their periodic orbits. See the equation *forcepen* above.

References: Arnold [1978] p. 117, [1980] p. 203; Guckenheimer & Holmes [1983] p. 29.

---

forceduf

Periodically forced Duffing's equation.

Type: 3D Differential.

Parameters: $a = 1.0$, $b = 0.2$, $c = 0.3$, $d = 1.0$.

Equations:

$$x_1' = x_2$$

$$x_2' = ax_1 - bx_2 - x_1^3 + c\cos(dx_3)$$

$$x_3' = 1.0 \qquad (mod\ \frac{2\pi}{d}).$$

This is the periodically forced oscillator of Duffing, with a cubic stiffness term to describe the hardening spring effect observed in many mechanical systems. Along with van der Pol's equation, it is one of the canonical examples in the theory of nonlinear oscillations. Duffing's equation is a nonautonomous system in $\mathbf{R}^2$. However, the appropriate phase space to use is $\mathbf{R}^2 \times S^1$ since the third variable, time, figures into the system as an argument of a periodic function. The remarkable sequence of bifurcations leading to apparently bounded non-periodic motion (strange attractor) can be seen through planar Poincare maps, using $x_3 = 3.0$, for example. See *Figures 7.19-20* for a sample non-periodic orbit and its Poincare map.

References: Chow & Hale [1982] p. 301; Guckenheimer & Holmes [1983] p. 82; Hale [1969] p. 195.

---

rossler

A not-so-strange strange attractor.

Type: 3D Differential.

Parameter: $a = 5.7$.

Equations:

$$x_1' = -(x_2 + x_3)$$

$$x_2' = x_1 + 0.2x_2 .$$

$$x_3' = 0.2 + x_3(x_1 - a) .$$

Rossler's equations provide a good example of the transition from simple to complicated behavior via a sequence of *period-doubling* bifurcations. For example, at the parameter value $a = 2.2$, there is an attracting limit cycle; see *Figure 7.21.* As the parameter a is increased, the period of this periodic orbit keeps doubling (see *Figure 7.22*) until $a \approx 4.2$, at which point a single solution fills up an attracting sheet, reminiscent of a mobius band. See also the equation *silnikov* and *Figure 7.25.*

Unlike the Lorenz equations, the equations of Rossler are artificially manufactured as an example of a strange attractor with a "simple" structure. A proof of the existence of such an attractor is still lacking, however.

References: Abraham & Shaw [1984]; Lichtenberg & Lieberman [1983] p. 386; Rossler [1976].

---

zeroim

Unfolding of zero plus imaginary eigenvalues.

Type: 3D Differential.

Parameters: $a = 2.015$, $b = 3.0$, $c = 0.25$, $d = 0.2$, $e = 0.0$.

Equations:

$$x_1' = (a - b)x_1 - cx_2 + x_1[x_3 + d(1.0 - x_3^2)]$$

$$x_2' = cx_1 + (a - b)x_2 + x_2[x_3 + d(1.0 - x_3^2)]$$

$$x_3' = ax_3 - (x_1^2 + x_2^2 + x_3^2) + ex_2^2x_3 .$$

These equations are a versal deformation around a degenerate equilibrium of a vector field whose linear part has a pair of pure imaginary and a zero eigenvalue. It is a bifurcation of codimension two, with the versal deformation parameters a and b.

This example, studied by Langford, is a good illustration of a *multiple bifurcation* arising from the coalescence of a Hopf bifurcation with a simple bifurcation of an equilibrium. To follow this interesting bifurcation, set $b = 3.0$, $c = 4.0$, $d = 0.2$, $e = 0.0$, and vary a in the approximate range from 0 to 2.5. When $a < 0$, the origin is stable. For small $a > 0$, the origin becomes unstable and a new stable equilibrium with $x_3 > 0$ appears. The new equilibrium undergoes a Hopf bifurcation at $a = 1.68$, giving rise to a stable limit cycle. At $a = 2.0$ the *Floquet multipliers* (eigenvalues of the Poincare map) of the periodic orbit are on the unit circle, but due to the higher order terms the limit cycle is still attracting. For $a > 2.0$, an attracting torus (see *Figures 7.23-4*) appears, and grows rapidly in diameter.

If the parameter e is varied in the range from 0 to 0.006, then the nonaxisymmetric term causes the torus to fold and stretch, leading to complicated dynamics.

References: Chow & Hale [1982] p. 462; Hale [1969] p. 118; Guckenheimer & Holmes [1983] p. 376; Langford [1982], [1985].

silnikov

A homoclinic orbit and horseshoes in three dimensions.

Type: 3D Differential.

Parameters: $a = 0.3375$, $b = 0.75$, $c = 0.633625$.

Equations:

$$x_1' = x_2$$

$$x_2' = x_3$$

$$x_3' = -x_2 - ax_3 + f(x_1), \qquad \textit{where}$$

$$f(x_1) = \begin{cases} 1.0 - bx_1 & \text{if } x_1 > 0.0 \\ 1.0 + cx_1 & \text{if } x_1 \leq 0.0\,. \end{cases}$$

Consider a system of differential equations in $\mathbf{R}^3$ which has an equilibrium point p such that the eigenvalues of the linearized equations at p are $\alpha \pm i\beta$, λ, with $|\alpha| < |\lambda|$ and $\beta \neq 0$. Suppose further that there is a *homoclinic orbit* for p -- that is, an orbit which tends to the equilibrium point p in both forward and reverse time. Then, according to Silnikov, there is a perturbation of such a differential equation with invariant sets containing transversal homoclinic orbits. The existence of such orbits implies the existence of Smale's *horseshoe*, which is a sure sign of complicated dynamics. (See the equation *henon* for a description of horseshoe.)

For the piecewise linear system above, studied by Arneodo et al., the hypotheses of the theorem of Silnikov can be checked. Indeed, when $a = 0.3375$ and $c = 0.633625$, there exists a value of the parameter b in the interval $[0.39, 0.4]$ for which the system has a homoclinic orbit asymptotic to the equilibrium point $(-1/c, 0, 0)$ in both forward and reverse times. The equation *silnikov* also has a visible strange attractor; see *Figure 7.25*.

Numerical evidence seems to suggest that the smooth version of this system with $f(x_1) = bx_1(1.0 - x_1)$ may have similar dynamics (see the equation *silnikov2*).

References: Arneodo et al. [1981], [1982]; Guckenheimer & Holmes [1983] p. 318; Silnikov [1965].

silnikov2 *A pair of Silnikov-like homoclinic orbits in three dimensions.*

Type: 3D Differential.

Parameters: $a = 0.4$, $b = 0.65$, $c = 0.0$, $d = 1.0$.

Equations:

$$x_1' = x_2$$

$$x_2' = x_3$$

$$x_3' = -ax_3 - x_2 + bx_1(1.0 - cx_1 - dx_1^2).$$

Numerical evidence suggests that this smooth system in three dimensions, studied by Arneodo et al., may have Silnikov-type equilibria (see the equation *silnikov*). For $c = 1.0$ and $d = 0.0$, this system is a smooth version of the equation *silnikov*. When $c = 0.0$ and $d = 1.0$, there seem to be values of the parameters a and b for which the system has *two* Silnikov-type equilibria; see ***Figure** 7.26*. For a discrete version of similar dynamics, you should consult the difference equation *act*.

References: Arneodo et al. [1981], [1982]; Guckenheimer & Holmes [1983] p. 318; Silnikov [1965].

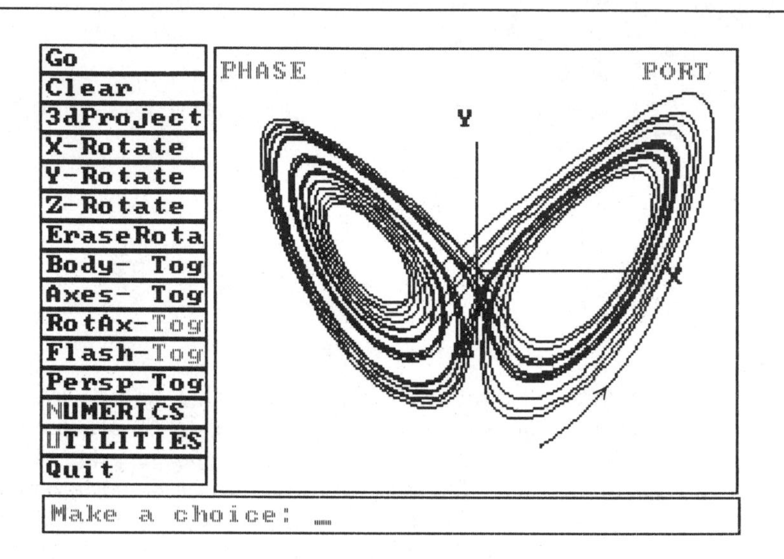

Figure 7.13. A nonperiodic orbit of the *lorenz* equations.

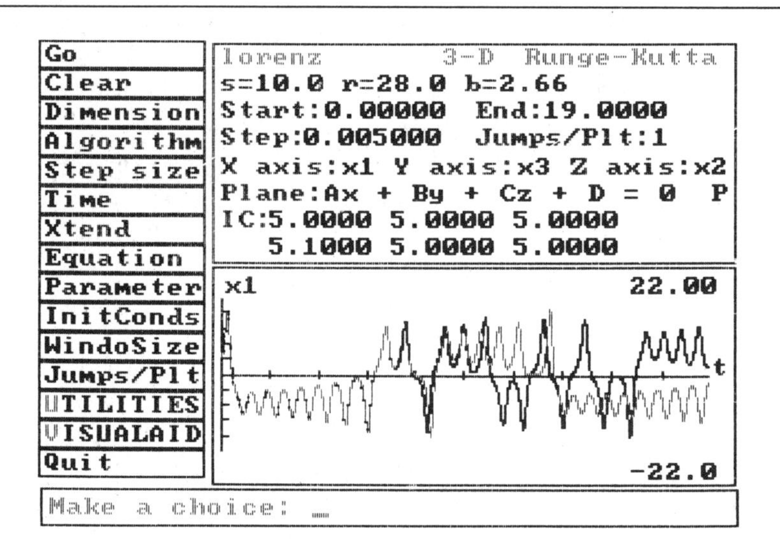

Figure 7.14. Sensitive dependence on initial conditions in the *lorenz* equations: two solutions starting close to each other diverge quickly.

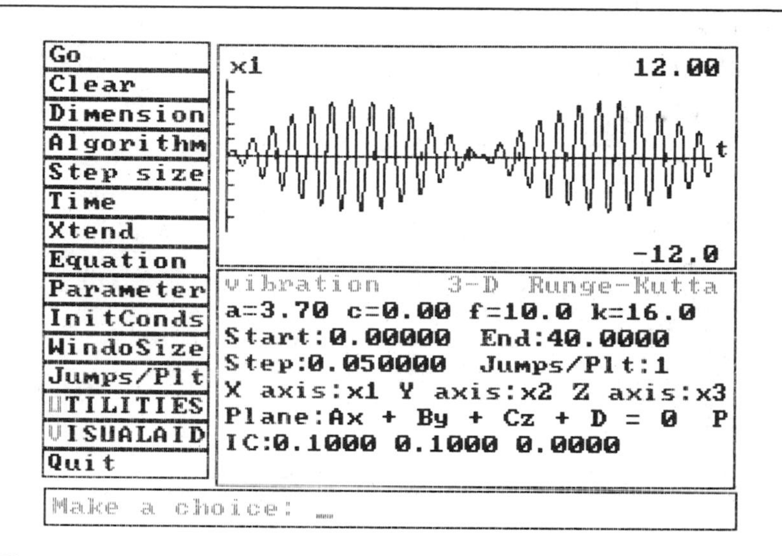

Figure 7.15. Phenomenon of beats, or wave packets, for the forced linear oscillator equation *vibration*.

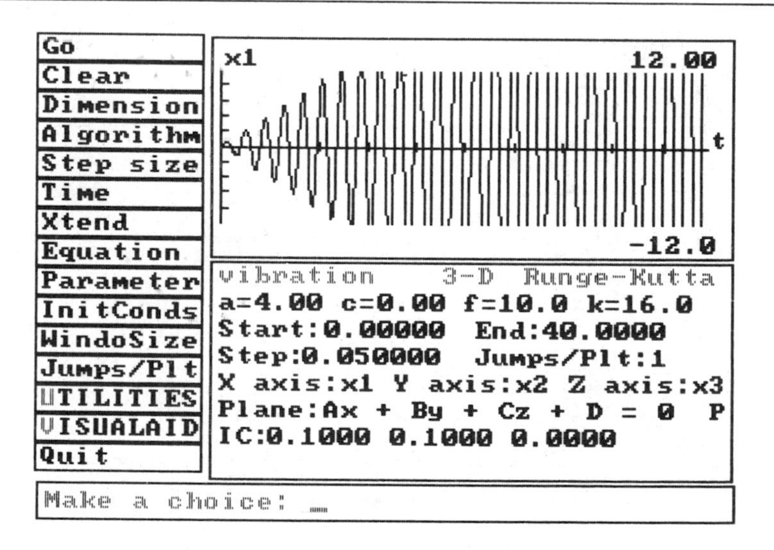

Figure 7.16. Phenomenon of resonance: a growing oscillatory solution of equation *vibration*.

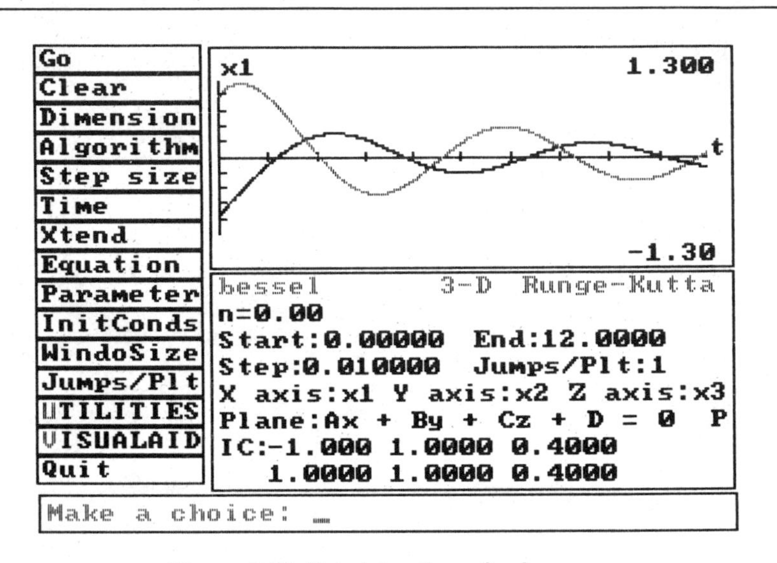

Figure 7.17. Bessel functions of order zero.

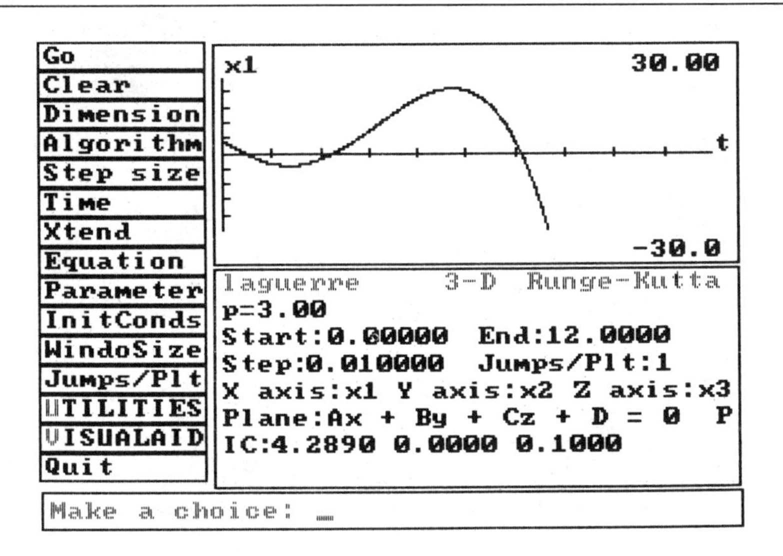

Figure 7.18. Laguerre polynomial of order three.

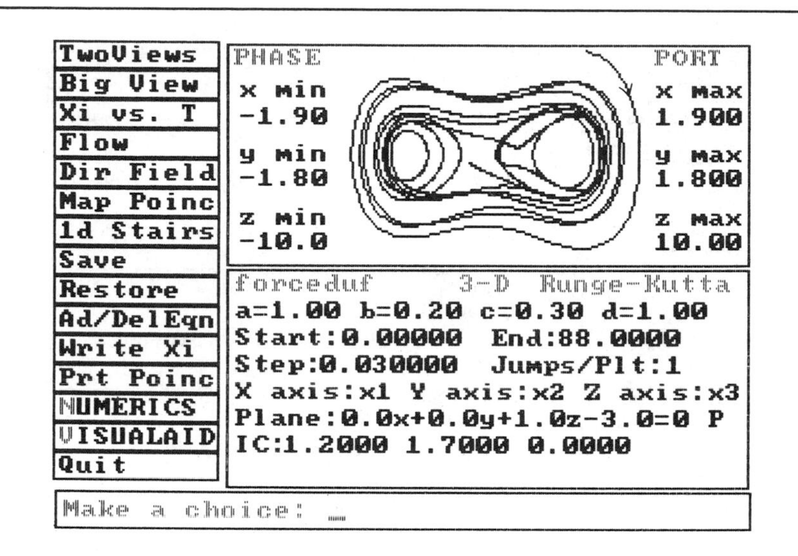

Figure 7.19. A nonperiodic orbit of the periodically forced Duffing's equation, *forceduf*.

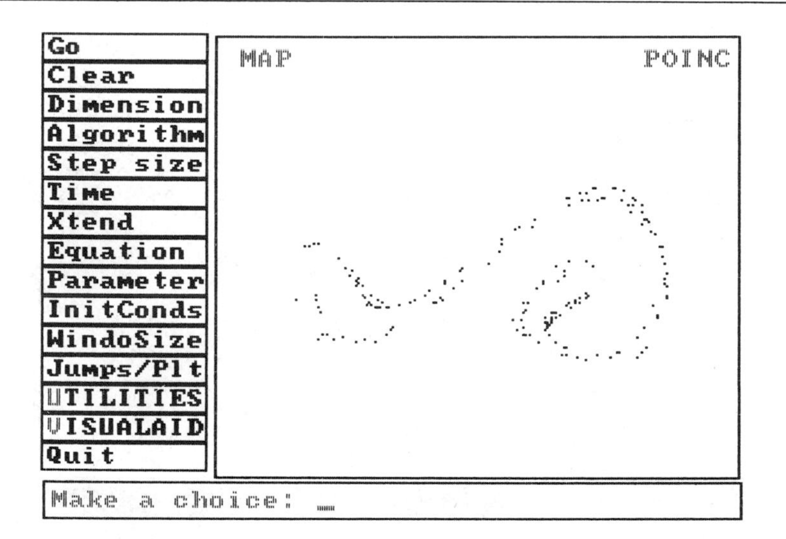

Figure 7.20. A planar Poincare map of the orbit in Figure 7.19 (notice the equation of the plane) of the periodically forced Duffing's equation (enlarged).

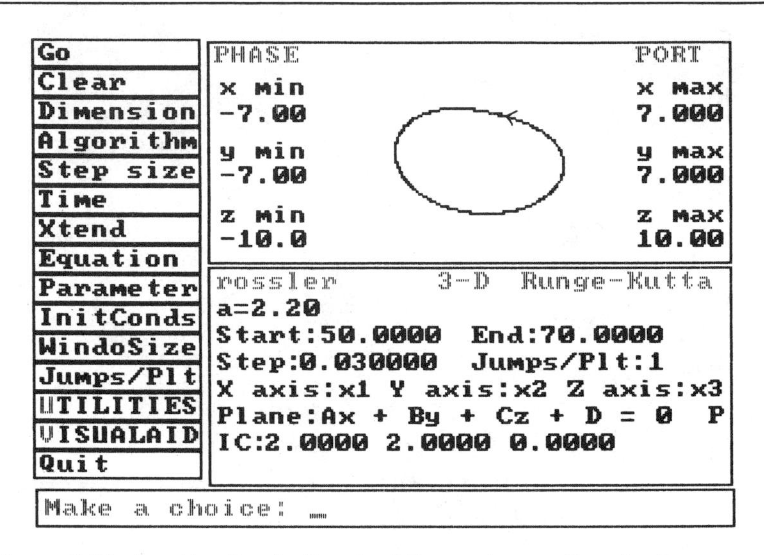

Figure 7.21. An attracting limit cycle of the *rossler* equations for the parameter value $a = 2.2$.

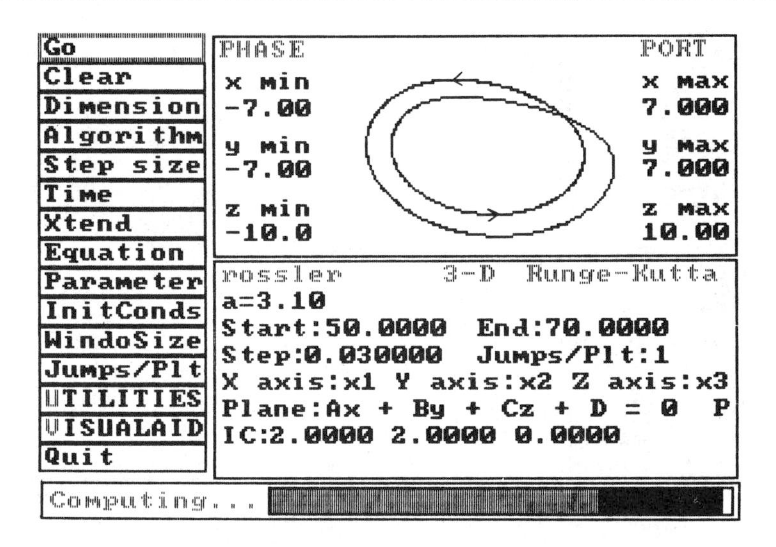

Figure 7.22. Period doubling in the *rossler* equations: for the parameter value $a = 3.1$, the period of the limit cycle in Figure 7.21 doubles.

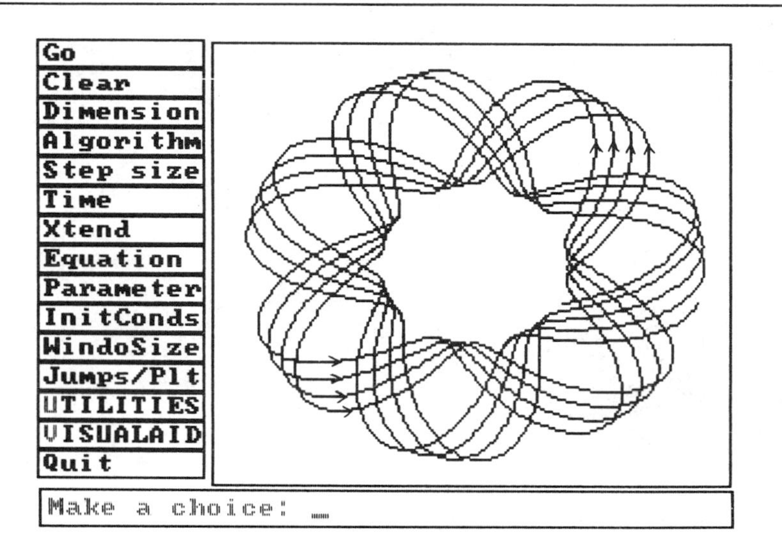

Figure 7.23. An attracting torus of the equation *zeroim* at the parameter value $e = 0.0$; see the setup view in Figure 7.24 for other parameter values, etc.

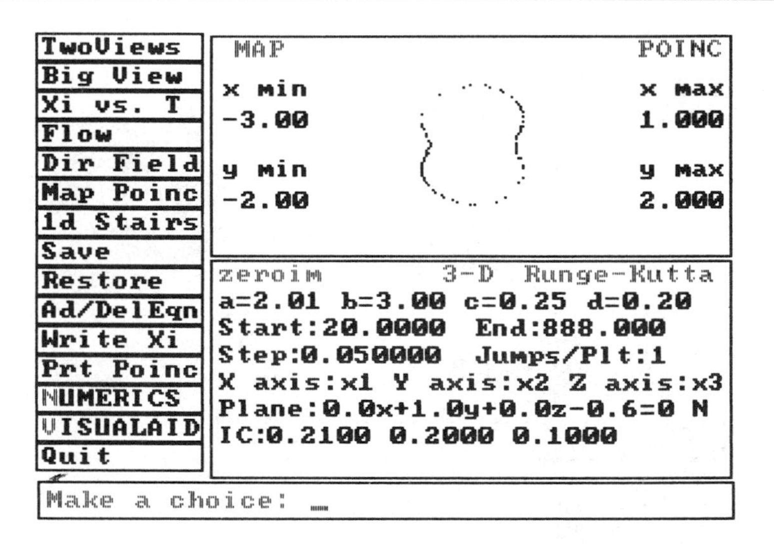

Figure 7.24. A planar Poincare map of equation *zeroim*, slicing through the invariant torus above (several solution curves are used). See the setup view for the equation of the plane and other settings.

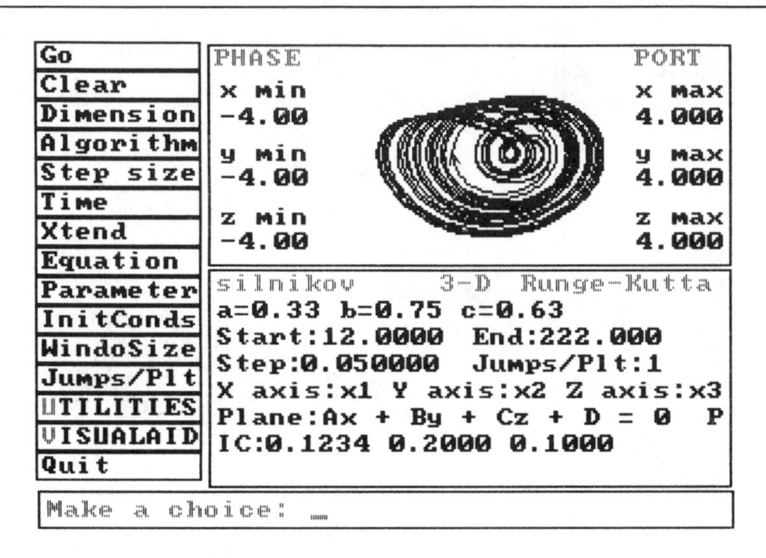

Figure 7.25. An apparent strange attractor of the equation *silnikov*.

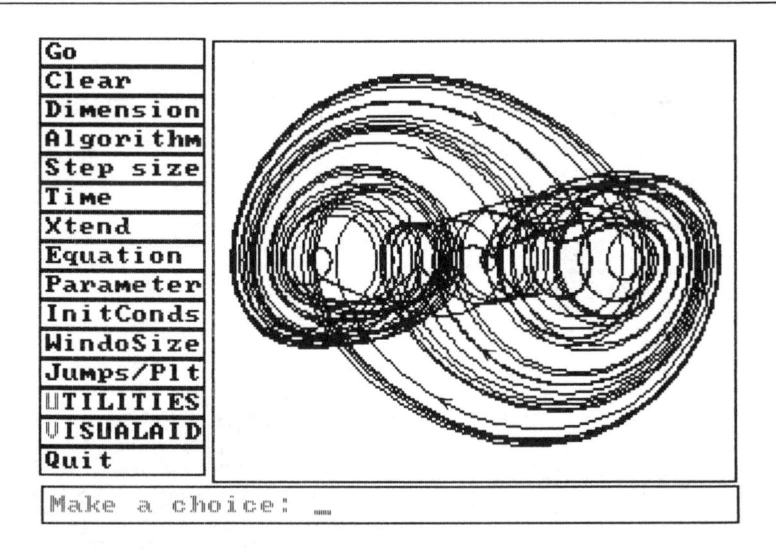

Figure 7.26. An apparent strange attractor of the equation *silnikov2* for the default values of the parameters; all other settings are as in *Figure 7.25*.

7.4. 4D Differential Equations

When studying equations in four dimensions or higher, you should use the *3dProject* entry on the *VISUALAID* menu to assign three of the variables to the x-, y-, and z-axes. Once solutions are projected into three dimensions, you can take advantage of all the usual graphical facilities.

---

harmoscil *A pair of linear harmonic oscillators.*

Type: 4D Differential.

Parameters: $a = 3.12121212$, $b = 2.1111111$.

Equations:

$$x_1' = ax_3$$

$$x_2' = bx_4$$

$$x_3' = -ax_1$$

$$x_4' = -bx_2 .$$

These equations describe the motion of two linear harmonic oscillators whose frequencies are a and b. This is a conservative system, with the Hamiltonian function

$$H = \frac{a}{2}(x_1^2 + x_3^2) + \frac{b}{2}(x_2^2 + x_4^2).$$

On the analytic side, these are linear equations with constant coefficients whose general solution can be easily written down. On the geometric side, however, they possess a surprisingly rich structure.

The best known pictures of this system are the orthographic projections of solution curves onto the (x_1, x_2)-plane (the configuration space) called Lissajous figures. The situation is much more complicated, however, than these pictures reveal. In fact, the orbits in general lie on two-dimensional tori in four-dimensional phase space, and the Lissajous figures are simply the "side views" of these tori. Orthographic projections of these tori into three-space (by dropping one of the coordinates) look like cylinders, as illustrated in Lesson 13. To see the "honest tori" on a two-dimensional screen is non-trivial, requiring sophisticated 4D graphics. For the full mathematical story of harmonic oscillators, consult the illustrated essay by Kocak et al.

References: Lessons 13, 14; Arnold [1978] p. 24, [1980] p. 160; Kocak et al. [1985].

---

kepler

Kepler and anisotropic Kepler problems.

Type: 4D Differential.

Parameters: $m = 1.0$, $p = 1.0$.

Equations:

$$x_1' = x_3$$

$$x_2' = x_4$$

$$x_3' = -\frac{mpx_1}{(px_1^2 + x_2^2)^{3/2}}$$

$$x_4' = -\frac{mx_2}{(px_1^2 + x_2^2)^{3/2}} .$$

When $p = 1.0$, these equations describe the planar motion of a particle that is attracted to the origin by a force which is inversely proportional to the square of the distance to the origin. This problem was investigated by Kepler and Newton, and is one of the landmarks in the history of calculus and mechanics. Its solutions are well known: all orbits lie on (possibly degenerate) conic sections; see *Figure 7.27.*

When $p > 1.0$, this system is called the ***anisotropic Kepler problem*** because the force is not directed toward the origin; rather, it is directed more toward the x_2-axis than the x_1-axis. The solutions of the anisotropic Kepler problem behave dramatically differently from those of the the Kepler problem; see *Figure 7.28.* In fact, numerical and theoretical evidence suggests that for $p > 9/8$, the anisotropic Kepler problem is an ***ergodic*** system on constant negative energy levels. (See Arnold & Avez [1968] for the use of ergodic theory in dynamical systems.)

Note: Since these equations are singular at the origin, their numerical simulations require special care.

References: Casasayas & Llibre [1984]; Devaney [1982]; Hirsch & Smale [1974] p. 14; Gutzwiller [1973]; Milnor [1983].

---

r3body

The restricted problem of three bodies on the plane.

Type: 4D Differential.

Parameter: $m = 0.5$.

Equations:

$$x_1' = x_3$$

$$x_2' = x_4$$

$$x_3' = 2x_4 + x_1 - (1-m)\frac{x_1 + m}{r_1^3} - m\,\frac{x_1 - 1 + m}{r_2^3}$$

$$x_4' = -2x_3 + x_2 - (1-m)\frac{x_2}{r_1^3} - m\,\frac{x_2}{r_2^3}, \quad \textit{where}$$

$$r_1 = [(x_1 + m)^2 + x_2^2]^{1/2},$$

$$r_2 = [(x_1 - 1 + m)^2 + x_2^2]^{1/2}.$$

Determining the motion of N point masses under the effect of their mutual inverse-square attraction is one of the central problems in celestial mechanics. Besides its practical importance, the so-called N-body problem has been a great source of challenge and inspiration to mathematicians.

The problem was essentially solved by Kepler for $N = 2$ (see the equation *kepler*). The next case, $N = 3$, has proven to be exceedingly difficult, and is still unsolved. However, attempts to solve a special case, called the *planar restricted three-body problem*, have enjoyed reasonable success.

Consider two primary bodies, N_1 and N_2, of mass m and $1 - m$, respectively, and a third body, N_3, of mass zero. The motions of N_1 and N_2 are given as a solution of a two-body problem because the third body does not influence the primaries. Assume further that the motion of N_3 takes place in the orbital plane of N_1 and N_2. The problem is then to describe the motion of N_3 on this plane.

In suitably normalized rotating coordinates, N_1 has a fixed position at $(-m, 0)$ and N_2 has a fixed position at $(1-m, 0)$. The motion of the third body, N_3, in the (x_1, x_2)-plane is given by the differential equations above.

These equations can be written as a Hamiltonian system by introducing the variables

$$q_1 = x_1, \quad q_2 = x_2, \quad p_1 = x_3 - x_2, \quad p_2 = x_4 + x_1,$$

and the Hamiltonian

$$H = \frac{1}{2}(p_1^2 + p_2^2) + p_1 q_2 - p_2 q_1 - \frac{1-m}{r_1} - \frac{m}{r_2}.$$

The *conserved quantity* (a function that is constant on the orbits) $C \equiv -2H$ is called the *Jacobi integral*, and in the original coordinates it is given by

$$C = x_1^2 + x_2^2 + \frac{2(1-m)}{r_1} + \frac{2m}{r_2} - x_3^2 - x_4^2.$$

There is no other known conserved quantity. (See the equation *henheile* for the implications of this!) For high values of the Jacobi integral (for example, 4.3), most orbits of the system are either periodic or quasi-periodic; see *Figure 7.29*. For lower values of C (for example, 3.3), however, there are orbits of N_3 which shuttle between the two primaries in a nonperiodic way; see *Figure 7.30*.

References: Henon [1983]; Siegel & Moser [1971] p. 256.

---

henheile *A non-integrable Hamiltonian by Henon & Heiles.*

Type: 4D Differential.

Parameters: $a = 1.0$, $b = 1.0$, $c = -1.0$.

Equations:

$$x_1' = x_3$$

$$x_2' = x_4$$

$$x_3' = -ax_1 - 2.0x_1x_2$$

$$x_4' = -bx_2 - x_1^2 - cx_2^2.$$

For the parameter values above, this is a landmark example in experimental dynamics, which was first studied by Henon and Heiles in conjunction with their work on the motion of a star in a cylindrical galaxy. It is a conservative system of two degrees of freedom with the Hamiltonian

$$H = \frac{1}{2}(ax_1^2 + bx_2^2 + x_3^2 + x_4^2) + x_1^2 x_2 + \frac{c}{3} x_2^3 .$$

If a Hamiltonian system with two degrees of freedom has another conserved quantity that is independent of the Hamiltonian, then almost all bounded orbits are either periodic or quasi-periodic and they lie on two-dimensional tori. Such systems are called *completely integrable* because one can almost integrate the differential equations explicitly (cf. Arnold [1978] p. 271). For example, the equations *harmoscil* and *kepler* are two elementary examples of completely integrable systems. However, if there is no second conserved quantity (the non-integrable case), then orbits can wander on three-dimensional constant energy surfaces, exhibiting very complicated behavior.

The system of Henon-Heiles is the most famous example of a non-integrable Hamiltonian. For small values of the Hamiltonian function, the system can be treated as a small perturbation of a completely integrable system; many of the orbits still lie on tori (KAM theory; cf. Arnold, Moser). For high energies, say $H = 1/6$, almost all the tori disintegrate and a single orbit can fill up an entire three-dimensional constant energy surface. Both of these cases can be seen experimentally by using Poincare maps. The non-integrability of the example of Henon-Heiles for high values of the Hamiltonian has finally been proven (cf. Churchill et al.).

The generalization of the Henon-Heiles system above is completely integrable for the following three sets of parameter values: $a = b$, $c = 1.0$; a and b arbitrary, $c = 6$; $b = 16a$, $c = 16$.

References: Aizawa & Saito [1972]; Arnold [1978] p. 405; Bountis et al. [1982]; Churchill et al. [1979]; Henon & Heiles [1964]; Moser [1968].

coplvdp1 *Two coupled van der Pols -- An attracting invariant torus.*

Type: 4D Differential.

Parameters: $b = 0.1$, $d = 0.1$, $e = 0.5$.

Equations:

$$x_1' = x_2$$

$$x_2' = -x_1 + b(x_3 - x_1) + e(1.0 - x_1^2)x_2$$

$$x_3' = x_4$$

$$x_4' = -x_3 - dx_3 + b(x_1 - x_3) + e(1.0 - x_3^2)x_4 .$$

This system is a pair of van der Pol oscillators coupled by weak linear interaction, b, with weak detuning, d. For $b = d = 0$, each oscillator has a globally attracting unique limit cycle (see the equation *vanderpol*); consequently, the system has a globally attracting torus. Under small perturbations such as the addition of weak coupling or detuning, the torus persists. However, the structure of orbits on the torus can change radically. For example, if $|b| < |d/2|$, then all the orbits on the torus are either periodic or almost periodic, with no attractors or repellers; see *Figure 7.31*. However, for $|b| > |d/2|$, there will be precisely two periodic orbits on the torus, one an attractor (see *Figure 7.32*) and the other a repeller (phenomenon of phase locking). During the transition from the first case to the second, a complex sequence of bifurcations can occur.

References: Guckenheimer & Holmes [1983] p. 59; Holmes & Rand [1980].

---

coplvdp2 *Two coupled van der Pols -- Two invariant tori.*

Type: 4D Differential.

Parameters: $a = 0.25$, $b = 1.0$, $c = 0.8$, $e = 0.00111$, $m = 0.2121$, $n = 0.1543212$.

Equations:

$$x_1' = x_2$$

$$x_2' = e(1.0 - x_1^2 - ax_3^2 - bx_1^2x_3^2)x_2 - m^2x_1$$

$$x_3' = x_4$$

$$x_4' = e(1.0 - cx_1^2 - x_3^2)x_4 - n^2x_3 .$$

This system of equations is a pair of van der Pol oscillators with nonlinear coupling. The main point of interest is the existence of invariant two-dimensional tori. For almost all parameter values, there is always one invariant torus whose stability type can be changed by a variation in b. However, with $b \neq 0.0$, there are also situations with two invariant tori, one of which is stable and the other of saddle type. The latter torus is not easy to locate numerically. The mechanism of the transition from the existence of two invariant tori to the existence of a single invariant torus is not well understood and deserves further study.

References: Baxter et al. [1972]; Chenciner [1983]; Hale [1963] p. 165.

---

couplosc *Coupled oscillators -- infinite period bifurcations and bistable behavior.*

Type: 4D Differential.

Parameters: $a = 1.0$, $b = 0.1$, $d = 0.0$.

Equations:

$$x_1' = x_1(a - x_1^2 - x_2^2) + bx_2 + d(x_3 - x_1 + x_4 - x_2)$$

$$x_2' = -bx_1 + x_2(a - x_1^2 - x_2^2) + d(x_3 - x_1 + x_4 - x_2)$$

$$x_3' = x_3(a - x_3^2 - x_4^2) + bx_4 + d(x_1 - x_3 + x_2 - x_4)$$

$$x_4' = -bx_3 + x_4(a - x_3^2 - x_4^2) + d(x_1 - x_3 + x_2 - x_4).$$

These equations describe a pair of identical limit cycle oscillators, each with natural frequency b, which are coupled by a linear diffusion path. The parameter a measures the amplitude of the limit cycle, while d measures the intensity of the coupling. For $d = 0.0$, there is an attracting normally hyperbolic invariant two-torus foliated by the individual limit cycles at arbitrary phase. For small enough $d > 0.0$, this torus persists, but is composed of a stable in-phase orbit ω_0, an unstable π-radians-out-of-phase orbit ω_π, and the unstable manifolds for ω_π. Although ω_π is unstable for small d, its restriction $\omega_\pi \mid \Pi$ to the invariant set

$$\Pi = \left\{ (x_1, x_2, x_3, x_4) \in \mathbf{R}^4 : x_1 + x_3 = 0, x_2 + x_4 = 0 \right\}$$

is an ellipse which is asymptotically stable in Π. For $b > a$, the diameter of $\omega_\pi \mid \Pi$ shrinks as d increases from zero, with ω_π disappearing in a Hopf bifurcation at the origin for $d = a/2$. For $b < a$, the Hopf bifurcation does not occur. Instead, $\omega_\pi \mid \Pi$ develops saddle-node equilibrium points when $d = b/2$ (infinite period bifurcation). For $d > b/2$, these saddle-nodes split into saddle-sink pairs, and $\omega_\pi \mid \Pi$ becomes the invariant circle consisting of heteroclinic orbits connecting the equilibrium points. Regardless of the value of $b > 0$, the system is bistable for sufficiently large $d > min(a/2, b/2)$. In particular, some initial conditions are attracted to the in-phase orbit ω_0, while other initial conditions are attracted either to an equilibrium point or a limit cycle which is not on Π; see *Figure 7.33*. There is also an open set of parameter values with $0 < d < min(a/2, b/2)$ where both of the orbits ω_0 and ω_π are asymptotically stable in $\mathbf{R}^4$. This set is rather hard to find numerically, however, since it is very small.

Reference: Aronson et al. [1985].

---

reson21 *Normal form of 2:1 resonance (non-hamiltonian).*

Type: 4D Differential.

Parameters: $a = 0.0$, $b = 0.0$, $c = 0.0$, $d = 0.0$, $m = 1.0$.

Equations:

$$x_1' = -mx_2 + a(x_1x_3 + x_2x_4) + b(x_3x_2 - x_1x_4)$$

$$x_2' = mx_1 + a(x_1x_4 - x_3x_2) + b(x_1x_3 + x_2x_4)$$

$$x_3' = -2mx_2 + c(x_1^2 - x_2^2) - 2dx_1x_2$$

$$x_4' = 2mx_1 + 2cx_1x_2 - d(x_1^2 - x_2^2).$$

This is an example of an unfolding of a degenerate equilibrium with a pair of pure imaginary eigenvalues which are in 2:1 resonance (the ratio of the eigenvalues is 2 to 1). The analysis of unfoldings of strong resonance cases, in which the two pure imaginary eigenvalues are related by a small integer, is still largely open.

References: Arnold [1983] p. 315; Chow & Hale [1982] p. 462; Guckenheimer & Holmes [1983] p. 413.

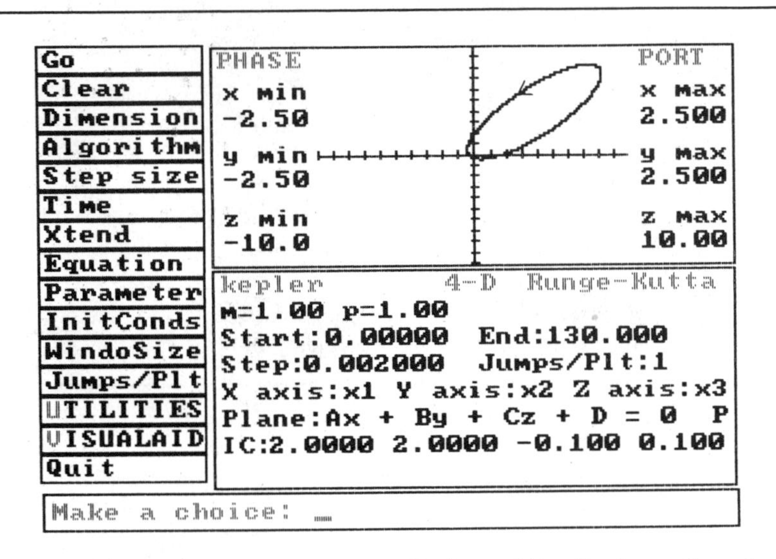

Figure 7.27. An elliptical orbit of the Kepler problem (in the configuration space).

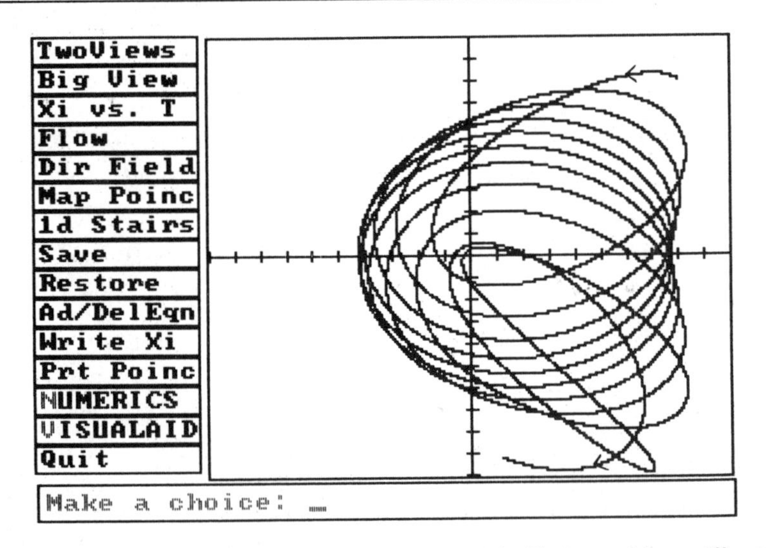

Figure 7.28. A nonperiodic orbit of the anisotropic Kepler problem. All settings are the same as in Figure 7.27, except that $p = 1.5$.

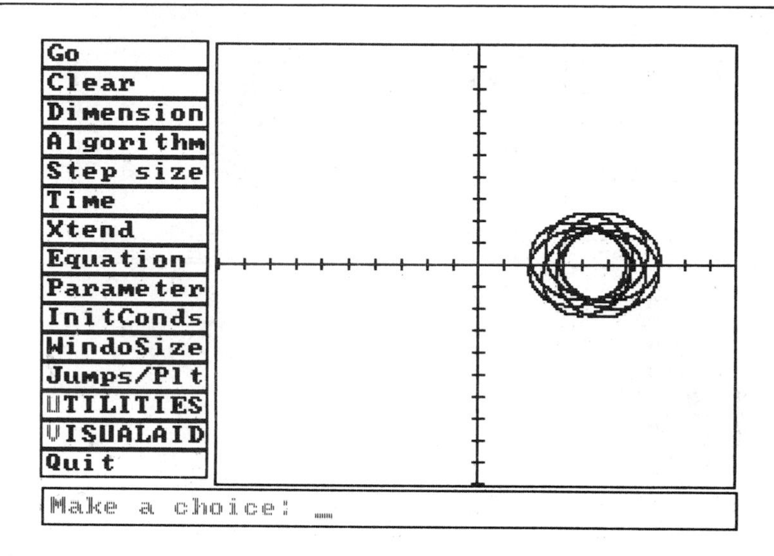

Figure 7.29. Capture by a primary in *r3body*: a quasi-periodic orbit of N_3, with the initial conditions $x_1 = 0.7$, $x_2 = 0.0$, $x_3 = -0.35$, $x_4 = 1.4$, and the step size 0.001.

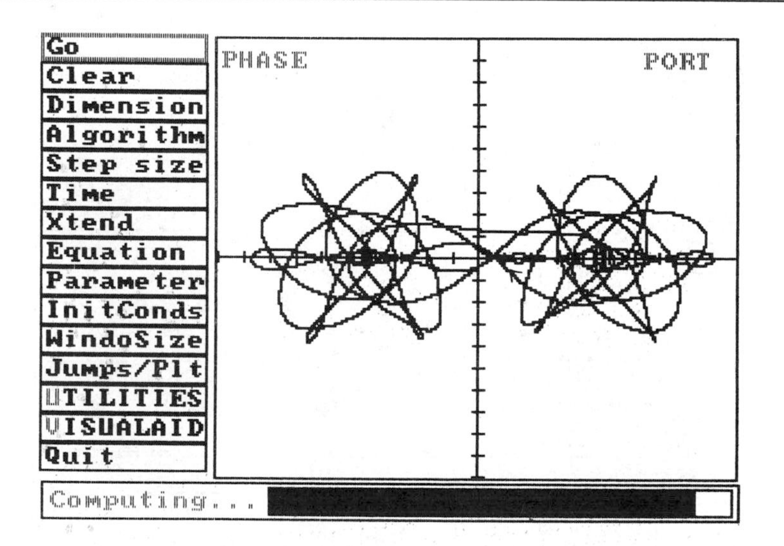

Figure 7.30. Shuttling between the two primaries in *r3body*: the orbit of N_3, with the initial conditions $x_1 = 0.7$, $x_2 = 0.0$, $x_3 = -0.35$, $x_4 = 1.6511$, and the step size 0.0002.

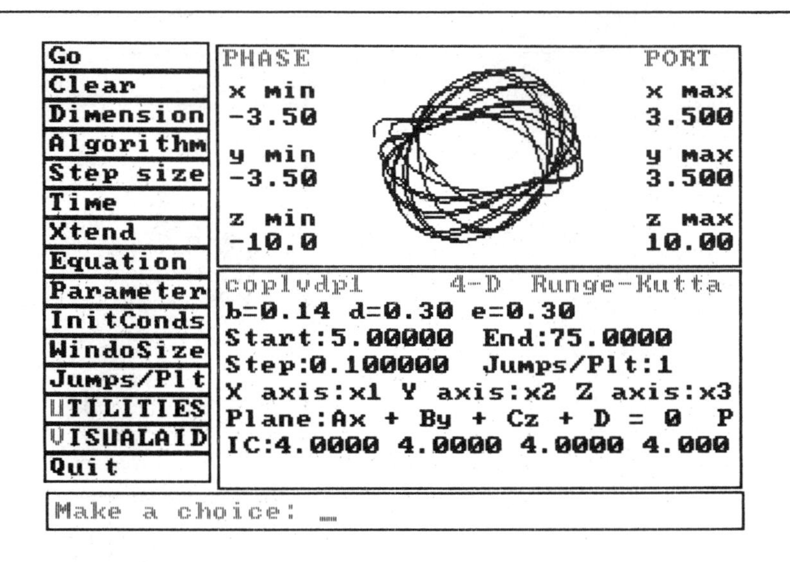

Figure 7.31. A quasi-periodic orbit on the unique invariant torus of the coupled van der Pol oscillators *coplvdp1* (with 30° rotation about the x-axis).

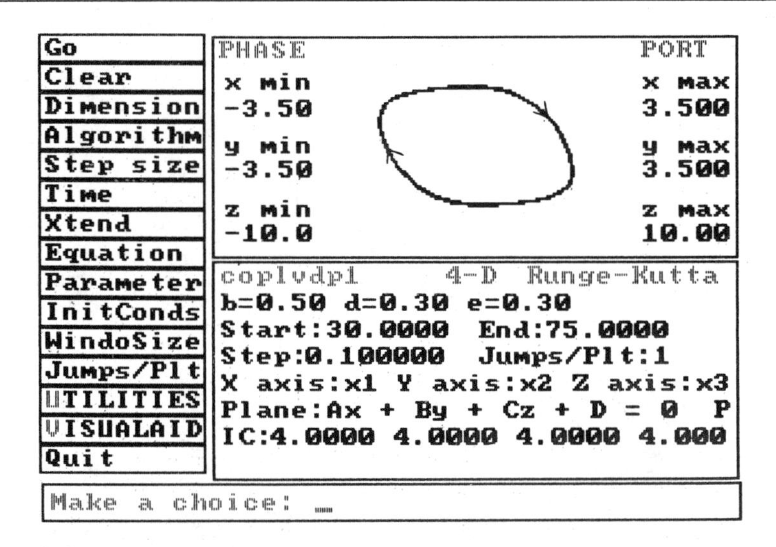

Figure 7.32. Attracting periodic orbit on the invariant torus of the coupled van der Pol oscillators *coplvdp1*: an example of 1:1 phase locking.

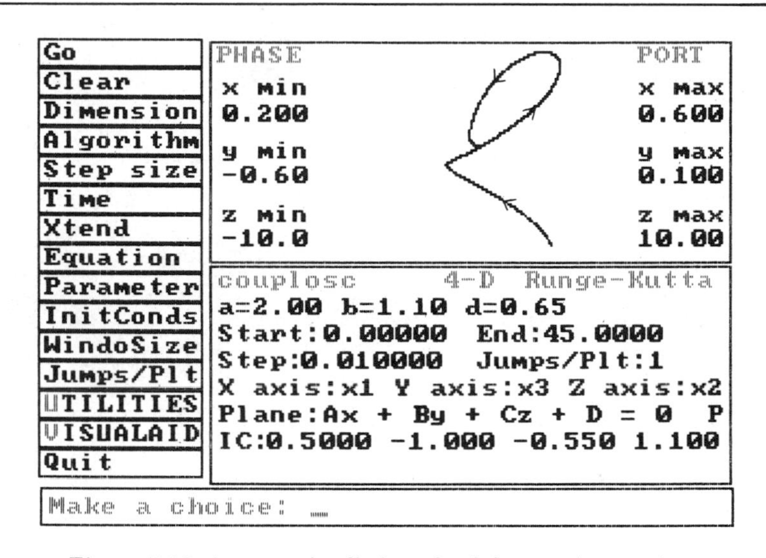

Figure 7.33. An attracting limit cycle of the equation *couplosc*.

Chapter 8

Difference Equations

This chapter is a catalogue of the difference equations stored in the library of PHASER. As it is similar to the catalogue of differential equations given in Chapter 7, you should now read the introduction to that chapter, if you have not already done so. Bear in mind, however, that in the formulas of difference equations x_i' denotes the next iterate of x_i, not its derivative.

8.1. 1D Difference Equations

When studying a one-dimensional difference equation, you should use the *1dStairs* view to see the graph of the equation and the iterates of an initial condition as a stair step diagram (see section 3.1 and Lesson 11).

---

logistic *The logistic map -- the one that started it all.*

Type: 1D Difference.

Parameter: $a = 3.41$.

Equation:

$$x_1' = a\ x_1 (1.0 - x_1)\ .$$

This one-dimensional difference equation is a simple model of a single population with limited carrying capacity (density dependence). It is called the *logistic* equation, and has long been known among ecologists. Until recently, however, its dynamic complexity has been realized by relatively few mathematicians. Nevertheless, it now appears that this ostensibly innocuous map has given rise to a whole new chapter in mathematics. I therefore made *logistic* the default equation for one-dimensional difference equations. See both section 3.1 and Lesson 11 of Chapter 5 for some of the elementary properties of this map as the parameter a is varied. Also, try $a = 3.83$.

By a simple coordinate transformation, the logistic map can be written as

$$x_1' = a - x_1^2 ,$$

which is the preferred form in the mathematical literature. (See the equation *discubic*, where several bifurcations of the map above are discussed.)

Note: The logistic map is the prototype for a class of interval maps with a single turning point, the so-called "one-hump" maps (see the equations *singer* and *tent*). The topological classification of such maps has been settled using the "kneading theory" of Milnor and Thurston. In addition, the spectacular "universal" properties of bifurcations discovered by Feigenbaum have been proved by Lanford.

References: Section 3.1, Lesson 11; Devaney [1985] p. 31; Guckenheimer & Holmes [1983] p. 307; Lanford [1980], [1982]; May [1976]; Maynard Smith [1968] p. 25; Whitley [1983].

---

dislin1d *General one-dimensional discrete linear equation.*

Type: 1D Difference.

Parameters: $a = 0.5$, $b = 0.0$.

Equation:

$$x_1' = ax_1 + b .$$

This is seemingly a trivial equation: if $|a| < 1$, then there is a unique globally attracting fixed point; if $|a| > 1$, then the unique fixed point is unstable (see section 3.1 and *Figure 3.1*). Yet, it is an important example to understand because the local dynamics of a nonlinear difference equation around a fixed point are generally reflected in its linearization (that is, in its derivative). For example, a fixed point of a one-dimensional nonlinear difference equation is stable if the derivative of the right-hand side of the equation at the fixed point is less than one in absolute value.

References: Section 3.1; Guckenheimer & Holmes [1983] p. 19.

---

discubic

General one-dimensional cubic difference equation.

Type: 1D Difference.

Parameters: $a = 0.0$, $b = -2.598$, $c = 0.0$, $d = 3.598$.

Equation:

$$x_1' = a + bx_1 + cx_1^2 + dx_1^3 .$$

Fixed points of difference equations become degenerate when the modulus of one eigenvalue is 1.0. In a one-parameter family of difference equations, degenerate fixed points, much like degenerate equilibria of differential equations, can undergo saddle-node, transcritical, pitchfork, and Hopf bifurcations. In addition, eigenvalues with −1.0 are associated with *flip* bifurcations (a stable fixed point becomes unstable and a new stable periodic orbit of period 2 appears). These bifurcations are also referred to as *period doubling* or *subharmonic* bifurcations, which have no analogues for equilibria.

All bifurcations except the Hopf bifurcation (see the equation *dellogis*) can be seen in this example. For instance, the case

$$x_1' = a - x_1^2$$

undergoes a saddle-node bifurcation at $x_1 = -0.5$, $a = -0.25$, and a flip bifurcation at $x_1 = 0.5$, $a = 0.75$ (see also the equation *logistic*).

The special case of *discubic*

$$x_1' = (1.0 - d)x_1 + dx_1^3$$

is used in genetics problems involving one locus with two alleles. Period-doubling bifurcations of this map accumulate at the value $d = 3.5980...$, beyond which there is "chaotic" behavior; see *Figures 8.1-2.*

References: Arnold [1983] p. 286; Devaney [1985] p. 79; Guckenheimer & Holmes [1983] p. 156; Rogers & Whitley [1983].

---

newton

Calculating square roots with Newton-Raphson.

Type: 1D Difference.

Parameter: $a = 2.0$.

Equation:

$$x_1' = 0.5(x_1 + \frac{a}{x_1}) .$$

This ancient example, which can be traced back at least as far as the Babylonians, illustrates how your pocket calculator computes the square root of a number. The difference equation above has a fixed point at $\sqrt{a}$. Since the fixed point is attracting, any positive initial guess near it quickly (quadratically) converges to the fixed point; see section 3.2. This example is a special case of a general scheme, called *Newton-Raphson's method*, for finding zeros of functions.

References: Section 3.2; Braun [1983] p. 87; Conte & deBoor [1980] p. 79; Saari & Urenko [1984].

---

tent

Piecewise linear version of the logistic map.

Type: 1D Difference.

Parameter: $a = 2.0$.

Equation:

$$x_1' = \begin{cases} ax_1 & \text{if } x_1 > 0.5 \\ a(1.0 - x_1) & \text{if } x_1 \leq 0.5 \end{cases}.$$

One simple example of a strange attractor for interval maps is the *logistic* map at the parameter value $a = 4.0$. Some of the properties of this strange attractor (for example, the existence of an "invariant measure") are simple consequences of the fact that the logistic map ($a = 4.0$) is *topologically conjugate* to the piecewise linear *tent* map ($a = 2.0$) via the homeomorphism

$$h(x) = \frac{2}{\pi} arcsin \sqrt{x} \;;$$

that is,

$$tent = (h) o (logistic) o (h^{-1}) .$$

Since these two maps are topologically conjugate for the parameter values indicated above, their dynamics must be qualitatively the same. However, it is interesting to observe that they behave rather differently in numerical simulations. For the parameter values above, the strange attractor is visible in the logistic map but not in the tent map, due to the binary nature of computer architecture. However, try the tent map with $a = 1.999999$. In fact, try any value of a between 1.0 and 2.0.

References: Devaney [1985] p. 52; Guckenheimer & Holmes [1983] p. 271; May [1976]; Ulam & von Neumann [1947].

---

singer

A one-hump map with two attractors.

Type: 1D Difference.

Parameter: $a = 1.0$.

Equation:

$$x_1' = a(7.86x_1 - 23.31x_1^2 + 28.75x_1^3 - 13.30x_1^4) .$$

The *singer* map has a unique maximum on the unit interval; visually it looks very much like the logistic or the tent map, the so-called "one-hump" maps. Initially, it was hoped that a one-hump map could be shown to have at most one stable periodic orbit. However, the *singer* map, for the parameter value $a = 1.0$, has both a stable fixed point and a stable periodic orbit of period 2. Since they are stable, it is easy to find these attractors by trying different initial conditions; see *Figures 8.3-4*.

References: Guckenheimer & Holmes [1983] p. 306; Singer [1978].

---

mod

Linear modulus map: A pseudo-random number generator.

Type: 1D Difference.

Parameters: $a = 1.0$, $b = 0.1414258$, $m = 1.0$.

Equation:

$$x_1' = a\, x_1 + b \qquad (mod\ \ m)\,.$$

This map is well known in several different areas of mathematics. For example, under the name *linear congruential method*, it is one of the most popular pseudo-random number generators. For the parameter values $a = 1.0$, $m = 1.0$, it is a rigid rotation of the circle with *rotation number* (average rotation) b, so that an orbit is uniformly distributed (ergodic) on the circle if b is irrational, and periodic otherwise. See the map *arnold* for a nonlinear perturbation of this map.

Reference: Knuth [1980] p. 9.

---

arnold

The standard circle map.

Type: 1D Difference.

Parameters: $a = 0.5$, $e = 0.1$.

Equation:

$$x_1' = x_1 + a + e\cos(2\pi x_1) \qquad (mod\ \ 1)\,.$$

If a Hamiltonian system with two degrees of freedom is close to a completely integrable system, then its bounded orbits usually lie on two-dimensional tori. Using Poincare maps, the dynamics on these tori can be reduced to diffeomorphisms of a circle. Indeed, Poincare initiated the study of circle diffeomorphisms in this context. Later, a rather deep analytical theory was developed by Denjoy, Arnold, and Herman. These results are also useful in non-hamiltonian systems where invariant tori are present; see, for example, the equations *coplvdp1* and *coplvdp2*.

For the parameter value $e = 0.0$, the map is a rigid rotation of the circle with rotation number a (see the equation *mod*). When $e \neq 0.0$, in the parameter space (a, e), the set of diffeomorphisms with rational rotation number p/q is bounded by a pair of smooth curves which come together at $e = 0.0$. Furthermore, these wedge-shaped regions, also called "Arnold tongues," get narrower with increasing q. The union of these sets is dense in the parameter space, but has small measure.

From the point of view of structural stability, one gets the impression that a generic diffeomorphism of the circle has rational rotation number. However, Arnold's two-parameter circle map above shows that this need not be the case in physical or numerical experiments. Stair step diagrams of this map for various values of the parameters are illustrated in *Figures 8.5-8*.

References: Arnold [1965], [1983] p. 108; Devaney [1985] p. 108; Guckenheimer & Holmes [1983] pp. 303, 350; Herman [1979]; Rand [1983].

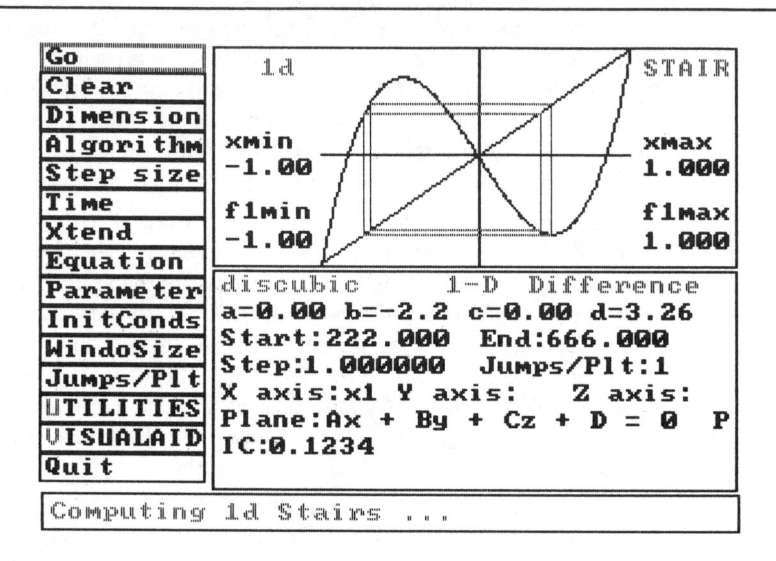

Figure 8.1. An attracting periodic orbit of period four of the map *discubic*.

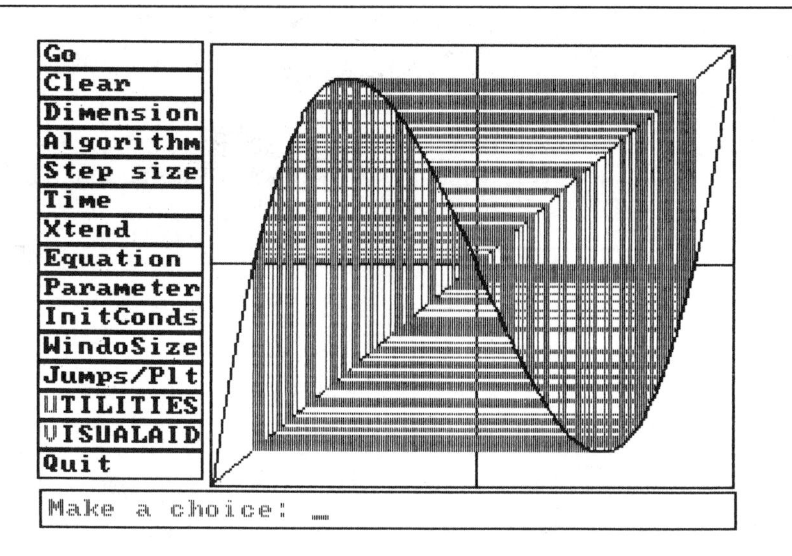

Figure 8.2. Beyond the period-doubling cascades in the map *discubic* (for the default values of the parameters).

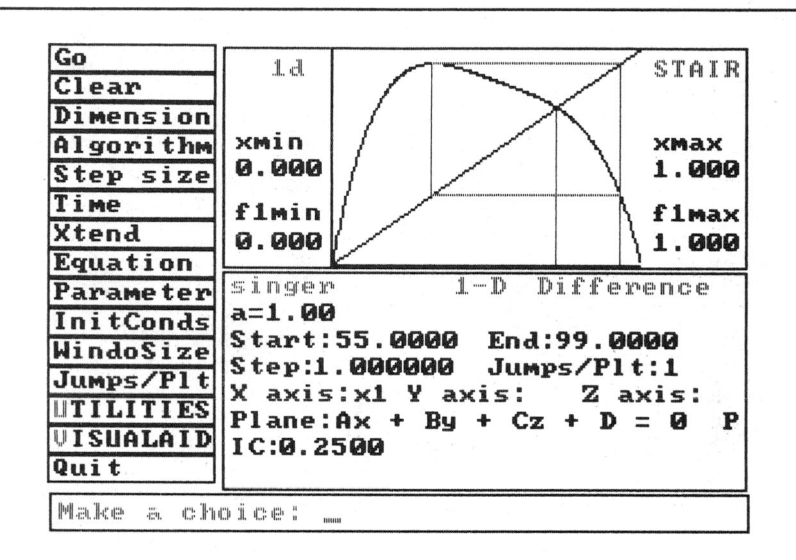

Figure 8.3. Graph of the *singer* map, and its coexisting stable fixed point (initial condition 0.2) and stable periodic orbit of period two (initial condition 0.25). To make the attractors more visible, the first few iterates of the orbits have been discarded.

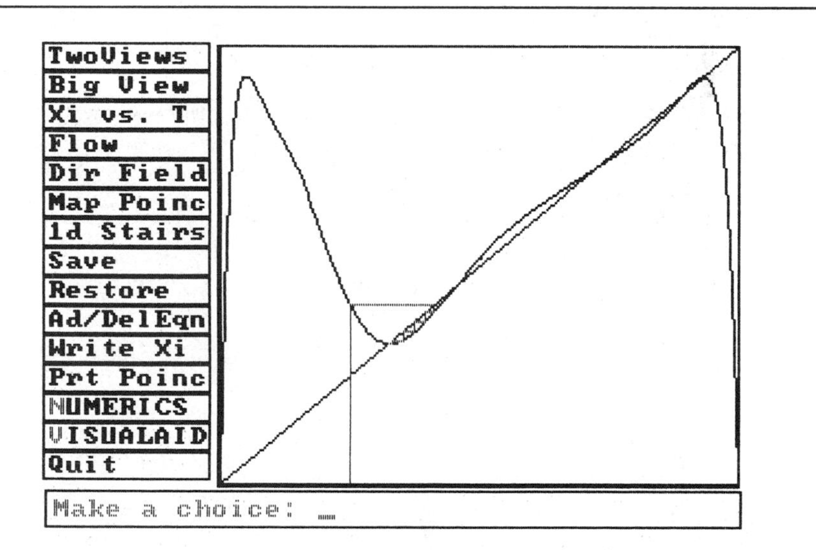

Figure 8.4. Graph of the second iterate of the *singer* map, and the orbit starting at 0.25.

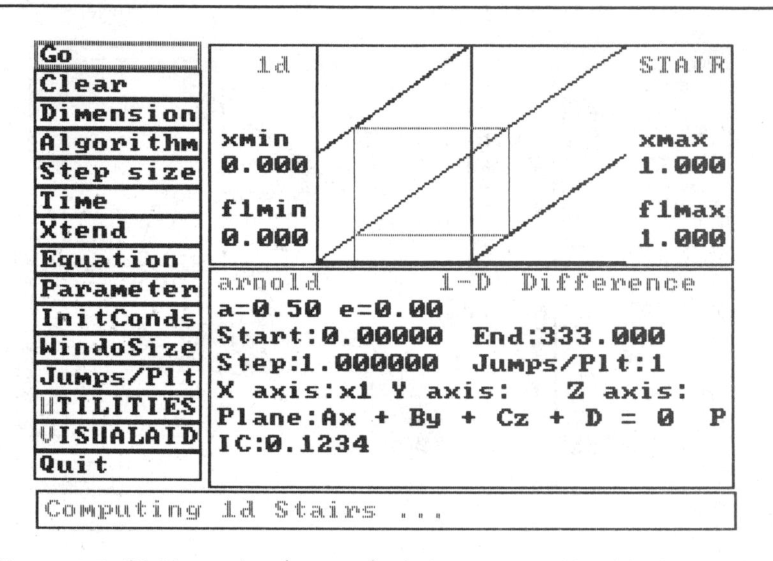

Figure 8.5. Rigid rotation (e = 0.0) of the map *arnold*, with the rotation number a = 1/2. (In the figures on this map ignore the dark vertical line.)

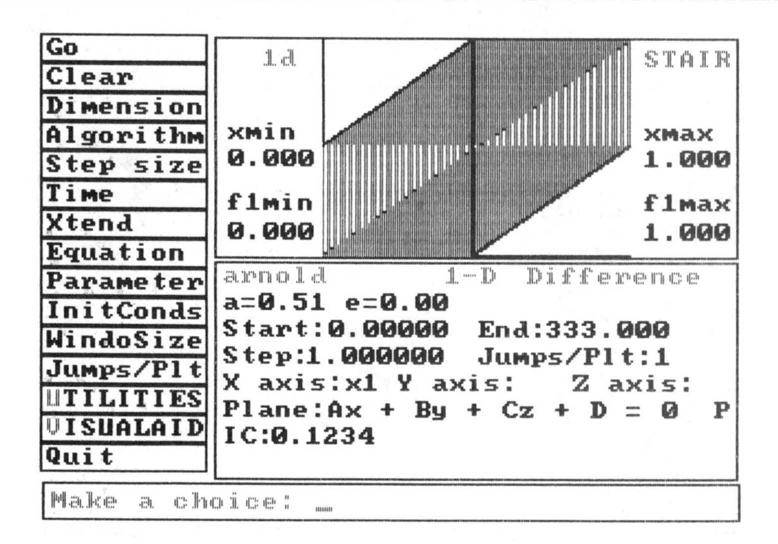

Figure 8.6. Rigid rotation (e = 0.0) of the map *arnold*, with the rotation number a = 51/100.

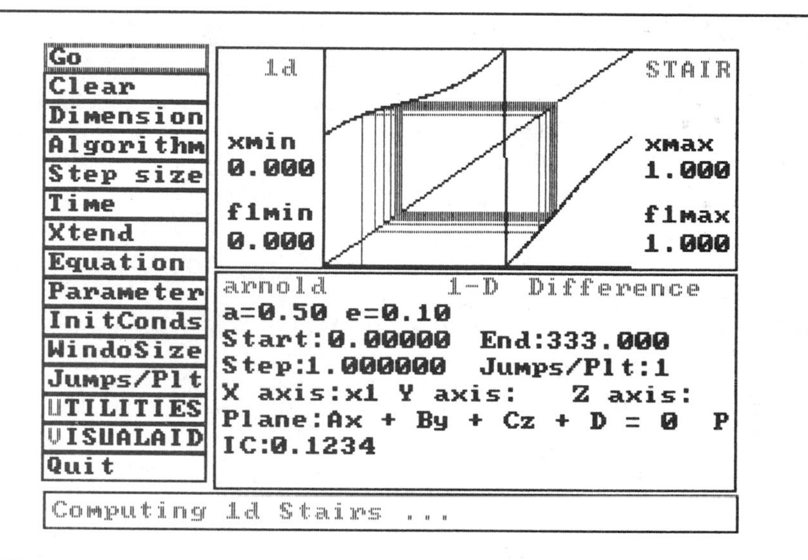

Figure 8.7. A stair step diagram of the map *arnold* for a set of parameter values inside the "Arnold tongue," with the rotation number 1/2 (compare with Figure 8.5).

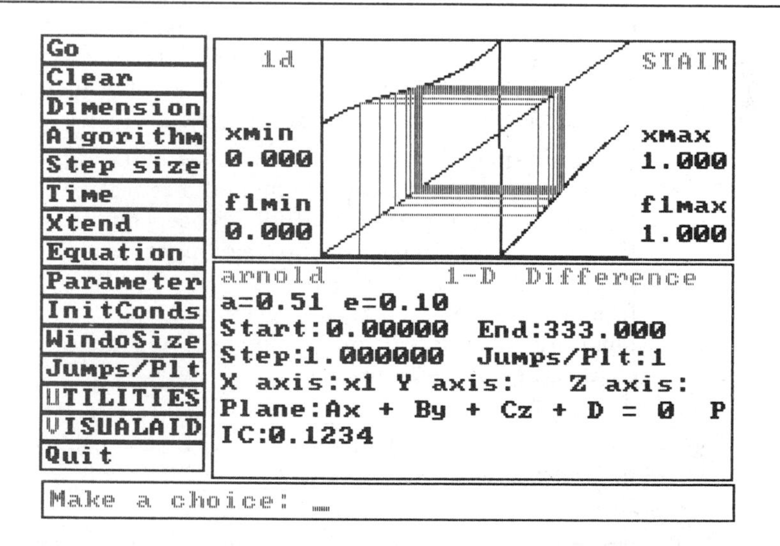

Figure 8.8. A stair step diagram of the map *arnold* for another set of parameter values in the "Arnold tongue," with the rotation number 1/2 (compare with Figure 8.6).

8.2. 2D Difference Equations

To follow the orbits of two-dimensional maps, you should activate the flasher on the VISUALAID menu. The *Jumps/Plt* is also useful in detecting periodic orbits when it is set to the expected period of the orbit.

---

dislin2d *General two-dimensional discrete linear system.*

Type: 2D Difference.

Parameters: $a = 0.0$, $b = 1.0$, $c = -1.0$, $d = 0.0$.

Equations:

$$x_1' = ax_1 + bx_2$$

$$x_2' = cx_1 + dx_2 .$$

As in the theory of differential equations, the chief importance of discrete linear systems stems from the fact that the phase portrait of a system of nonlinear difference equations around an equilibrium point looks qualitatively like that of its linearization (except possibly in the presence of eigenvalues with modulus one).

From an analytical point of view, it is quite easy to solve these equations using eigenvalue and eigenvector methods. However, phase portraits of difference equations, unlike those of differential equations, can be difficult to recognize, as illustrated in *Figures 8.9-10.* You should therefore investigate, by varying the parameters, all possible qualitatively different phase portraits of this system, as classified in the references.

References: Devaney [1985] p. 170; Guckenheimer & Holmes [1983] p. 19.

---

gauss *Fast computation of elliptic integrals.*

Type: 2D Difference.

Parameter: None.

Equations:

$$x_1' = \frac{1}{2}(x_1 + x_2)$$

$$x_2' = \sqrt{x_1 x_2}\,.$$

This example is the arithmetic-geometric mean iteration of Gauss and Legendre for quadratically approximating the following complete elliptic integral of the first kind:

$$I(a\,,b\,) = \int_0^{\pi/2} \frac{d\,\theta}{\sqrt{a^2\cos^2\theta + b^2\sin^2\theta}}\,.$$

For the initial conditions $x_1^0 = a$ and $x_2^0 = b$, both x_1^n and x_2^n converge quadratically, as $n \to +\infty$, to the value of the integral $I(a\,,b\,)$.

Reference: Borwein & Borwein [1984].

---

quad1

A quadratic map on the plane.

Type: 2D Difference.

Parameters: $a = 0.3$, $b = 1.4$.

Equations:

$$x_1' = x_2$$

$$x_2' = ax_1 + b - x_2^2\,.$$

Apart from a simple change of coordinates, this quadratic map of the plane is essentially the map of Henon with constant Jacobian (see the equation *henon*). Some of its complicated dynamics can be understood in terms of one-dimensional maps. When $a = 0.0$, *quad1* becomes singular and collapses the entire plane onto a parabola on which the dynamics are governed by the *logistic* map. Many of the known results for the logistic map can be extended to the non-singular case, $a \neq 0$. However, there are additional complicated dynamical behaviors which cannot be captured by this one-dimensional analysis. For example, the map *quad1* contains *homoclinic bifurcations* in which the stable and the unstable manifolds of a saddle point become tangent and then intersect transversally. This implies the possible existence of infinitely many stable periodic orbits for the parameter values near the homoclinic tangency.

References: Newhouse [1980]; Whitley [1982].

---

quad2

A quadratic map exhibiting a Hopf bifurcation.

Type: 2D Difference.

Parameters: $a = -1.6$, $b = 1.8$.

Equations:

$$x_1' = x_2$$

$$x_2' = ax_2 + b - x_1^2 .$$

This map is included in the library because it exhibits a Hopf bifurcation (see *Figures 8.11-12*), a phenomenon which is present in neither the Henon map, nor the map *quad1*. In fact, *quad2* displays examples of all the weak and strong resonances (presence of eigenvalues that are roots of unity) associated with a Hopf bifurcation. For a related map, see also the equation ***dispprey***.

References: Guckenheimer & Holmes [1983] pp. 160, 165; Whitley [1982].

---

henon

A quadratic map with a strange attractor.

Type: 2D Difference.

Parameters: $a = 1.4$, $b = 0.3$.

Equations:

$$x_1' = 1.0 + x_2 - ax_1^2$$

$$x_2' = bx_1 .$$

This is the planar diffeomorphism of Henon, which appears to have a strange attractor whose local structure suggests a Cantor set cross an interval. The attractor is essentially indistinguishable from the unstable manifold of one of the saddle points. The Henon map is one of the prototypical examples of complicated dynamics, and has been studied intensively by both mathematicians and scientists.

It was shown by Devaney and Nitecki that for certain parameter values (which do not include those used by Henon), the Henon map has a Smale *horseshoe* -- a sure sign of complicated dynamics. This is a hyperbolic invariant set on which the map is topologically equivalent to the shift automorphism of symbolic dynamics. That is, one can find two regions of the plane, and also orbits of the map which visit these regions in any preassigned order. Unfortunately, the horseshoe is not an attractor; in the Henon map, points near this invariant set escape to infinity. So, what you see on the screen (*Figures 8.13-14*) still awaits a mathematical proof.

For variations on the Henon map, see the equations *lozi* and *quad1*.

References: Curry [1979]; Devaney [1985] p. 248; Devaney & Nitecki [1979]; Henon [1976]; Guckenheimer & Holmes [1983] p. 245; Whitley [1983].

---

lozi *Piecewise linear version of the Henon map.*

Type: 2D Difference.

Parameters: $a = 1.7$, $b = 0.5$.

Equations:

$$x_1' = 1.0 + x_2 - a \mid x_1 \mid$$

$$x_2' = bx_1 .$$

This is a piecewise linear map on the plane, possessing dynamical behavior similar to that seen in the map of Henon. However, while it has been proved by Misiurewicz that the Lozi map does have a hyperbolic strange attractor, no such proof exists as yet in the case of the Henon map.

References: Devaney [1985] p. 211; Guckenheimer & Holmes [1983] p. 267; Lozi [1978]; Misiurewicz [1980].

---

dellogis

Two-parameter delayed logistic map.

Type: 2D Difference.

Parameters: $a = 2.16$, $b = 0.0$.

Equations:

$$x_1{}' = x_2 + bx_1$$

$$x_2{}' = ax_2(1.0 - x_1)\ .$$

This is the discrete logistic model of a single population, except that the nonlinear term regulating the population growth contains a time delay after one generation. Here a is the biologically important parameter (intrinsic growth rate), and b is added for theoretical reasons (see section 3.3).

For each set of fixed values of the parameters a and b, the corresponding map has a fixed point in the first quadrant. In the neighborhood of this point, the map is a diffeomorphism. For certain parameter values (e.g., $a = 2.0$, $b = 0.0$) there is a Hopf bifurcation to an invariant circle, one which is smooth for parameter values close to the bifurcation values. Computer experiments show, however, that the corresponding invariant set fails to be even topologically a circle for parameter values far from the bifurcation values. You should consult Aronson et al. for an impressive attempt to elucidate, using a blend of computer experiments and mathematical theory, some of the mechanisms involved in this loss of smoothness and alteration of topological type.

References: Aronson et al. [1982]; Guckenheimer & Holmes [1983] p. 163; Maynard Smith [1968] p. 23; Pounder & Rogers [1980].

dispprey

Discrete predator prey model.

Type: 2D Difference.

Parameters: $a = 3.6545$, $b = 0.31$.

Equations:

$$x_1' = ax_1(1.0 - x_1) - x_1x_2$$

$$x_2' = \frac{1.0}{b}x_1x_2 \ .$$

The system of difference equations above is a discrete model of interactions between two species: a prey, x_1, and a predator, x_2. In the absence of the predator, the population of the prey is governed by the logistic equation. Furthermore, it is assumed that each predator kills a number of prey proportional to the abundance of that prey. For the growth rate of the predator, the assumption is that the number of offspring produced by each predator is proportional to the number of prey it kills.

As the parameters are varied, these equations undergo a remarkable sequence of bifurcations, ranging from stable fixed points to strange attractors; see *Figures 8.15-16.* For a related map, see the equation *quad2*.

Reference: Maynard Smith [1968] p. 27.

---

bounball

Bouncing ball on a periodically vibrating table.

Type: 2D Difference.

Parameters: $a = 0.8$, $b = 10$.

Equations:

$$x_1' = x_1 + x_2 \qquad (mod \ 2\pi)$$

$$x_2' = ax_2 - b \ cos\,(x_1 + x_2) \ .$$

This two-dimensional mapping provides a model for repeated impacts of a ball on a heavy, sinusoidally vibrating table. Here a is the coefficient of dissipation, and b is the forcing amplitude. Notice that because of the periodic motion of the table, the phase space is taken to be the cylinder $S^1 \times \mathbf{R}$.

One of the distinguishing features of the bouncing ball problem is that it provides an example of a mechanical system whose analysis leads directly to difference equations, bypassing any consideration of differential equations and Poincare maps. Although maps are much easier to simulate numerically than differential equations, the mathematical analysis of *bounball* turns out to be quite difficult: there are many families of periodic orbits, and also complicated invariant sets like Smale's horseshoe; see *Figure 8.17.*

Reference: Guckenheimer & Holmes [1983] p. 102.

---

anosov

Ergodic toral automorphism of Anosov.

Type: 2D Difference.

Parameters: $a = 1.0$, $b = 1.0$, $c = 1.0$, $d = 2.0$.

Equations:

$$x_1' = ax_1 + bx_2 \qquad (mod\ 1)$$

$$x_2' = cx_1 + dx_2 \qquad (mod\ 1)\ .$$

For the parameter values above, this is a linear area-preserving map ($|\ determinant\ | = 1$), except that the phase space is the two-dimensional torus obtained by identifying points whose coordinates differ by integers. All orbits therefore remain inside the unit square. The origin is a fixed point whose unstable manifold is dense on the unit square. Consequently, a single orbit, unless periodic, eventually fills up the entire square. You should run two orbits simultaneously to observe their uniform mixing; see *Figure 8.18.*

This automorphism of the torus is an Anosov system, and is therefore structurally stable (in the C^1-topology). That is, every automorphism sufficiently close to ***anosov*** together with its derivative is conjugate to ***anosov*** by an homeomorphism.

References: Arnold [1983] p. 122; Arnold & Avez [1968] p. 5; Devaney [1985] p. 189; Guckenheimer & Holmes [1983] p. 20.

---

cremona *An area-preserving quadratic (standard) map.*

Type: 2D Difference.

Parameter: $a = 1.32843$.

Equations:

$$x_1' = x_1 \, cos\,(a) - (x_2 - x_1^2)\, sin\,(a)$$

$$x_2' = x_1 \, sin\,(a) + (x_2 - x_1^2)\, cos\,(a)\,.$$

This is an invertible (see the equation *icremona* below) area-preserving map, which can be thought of as a model Poincare map of a four-dimensional system of Hamiltonian differential equations. The Cremona map possesses very complicated dynamics reminiscent of non-integrable Hamiltonian systems.

This map has two fixed points, one elliptic (the origin) and the other hyperbolic, as well as periodic orbits of different periods. Near the origin is a regular structure consisting of concentric invariant circles surrounding the elliptic fixed point. As we move outwards, the invariant circles become distorted and a chain of islands appear. Finally, we reach a "chaotic" region where there are no more invariant curves, and where the orbits begin to fill two dimensional regions; see *Figure 8.19*. The fine details of the picture are much more complicated, however: each individual island is a miniature replica of the larger structure, with an infinite hierarchy of islands possessing similar local structure.

In addition to the Cremona map, many other area-preserving maps have been studied numerically. In most cases the features outlined above seem to prevail (see the equation *gingerman*). A careful mathematical analysis of the dynamics of these maps is quite difficult, but the general features can be explained by the works of Kolmogorov, Arnold, and especially Moser.

References: Henon [1969], [1983]; Moser [1968].

---

icremona *The inverse of the cremona map.*

Type: 2D Difference.

Parameter: $a = 1.32843$.

Equations:

$$x_1' = x_1\cos(a) + x_2\sin(a)$$

$$x_2' = -x_1\sin(a) + x_2\cos(a) + (x_1\cos(a) + x_2\sin(a))^2 .$$

This is the inverse of the *cremona* mapping described above. In discrete dynamical systems, one cannot simply take a "negative step size" to follow orbits backward in time. It is for this reason that this mapping has been included in the library.

References: Henon [1969], [1983]; Moser [1968].

---

gingerman *A piecewise linear area-preserving map.*

Type: 2D Difference.

Parameter: None.

Equations:

$$x_1' = 1.0 - x_2 + |x_1|$$

$$x_2' = x_1 .$$

An elliptic fixed point of a *smooth* area-preserving map is surrounded by infinitely many invariant circles on which the mapping is an irrational rotation; between these circles are annular zones of instability. This is known as the *Moser twist theorem.* The *gingerman* map is a *piecewise linear* map exhibiting such behavior around the elliptic fixed point (1, 1). You should experiment with different initial conditions to observe this fascinating phenomenon (see *Figure 8.20*), and compare your findings with the dynamics of the *cremona* map.

In general, theoretical analysis tends to be easier for piecewise linear maps than for smooth ones. Two notable examples are the equations *tent* and *lozi*. The *gingerman* is yet another instance of this phenomenon.

Reference: Devaney [1984].

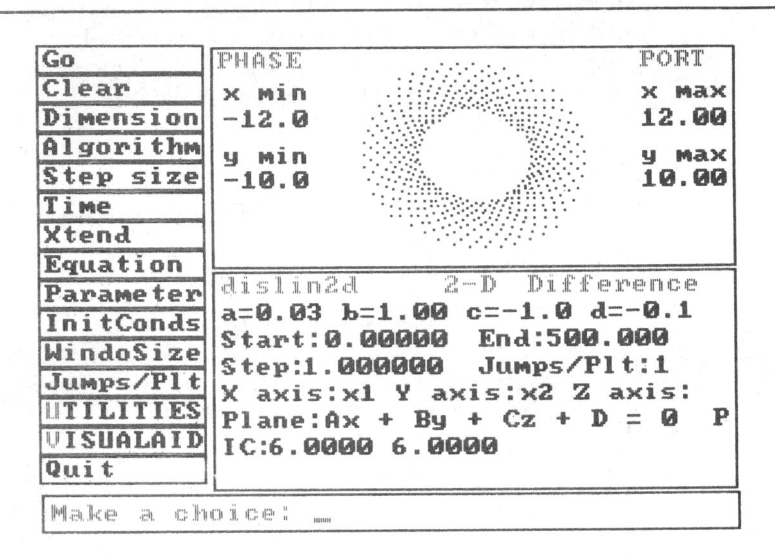

Figure 8.9. An orbit (single initial condition!), which is a spiral sink, of the linear difference equation *dislin2d*.

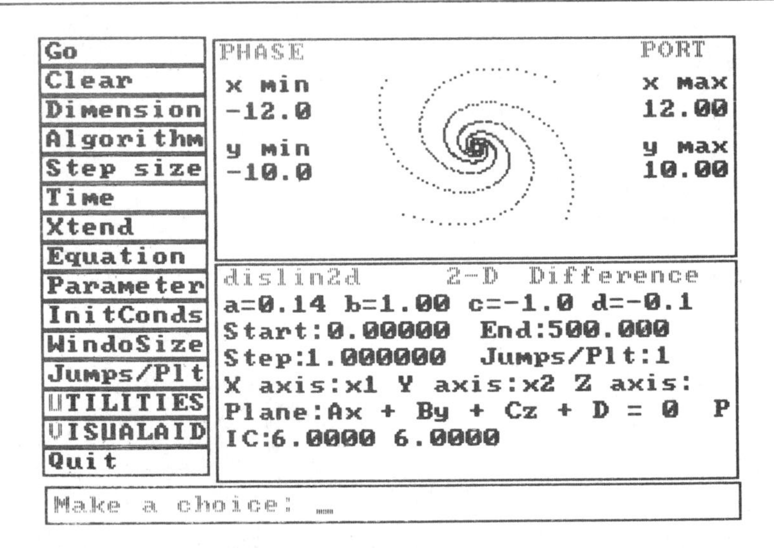

Figure 8.10. Another orbit (single initial condition!) of the map *dislin2d*; compare with *Figure 8.9*.

```
Go            PHASE                                   PORT
Clear         x min                                   x max
Dimension     -0.50                                   1.500
Algorithm     y min                                   y max
Step size     -0.50                                   1.500
Time
Xtend
Equation
Parameter     quad2            2-D   Difference
InitConds     a=0.32 b=0.72
WindoSize     Start:0.00000   End:444.000
Jumps/Plt     Step:1.000000   Jumps/Plt:1
UTILITIES     X axis:x1 Y axis:x2 Z axis:
VISUALAID     Plane:Ax + By + Cz + D = 0    P
Quit          IC:0.5500 0.5540
Make a choice: _
```

Figure 8.11. An invariant circle of the map *quad2* after a supercritical Hopf bifurcation.

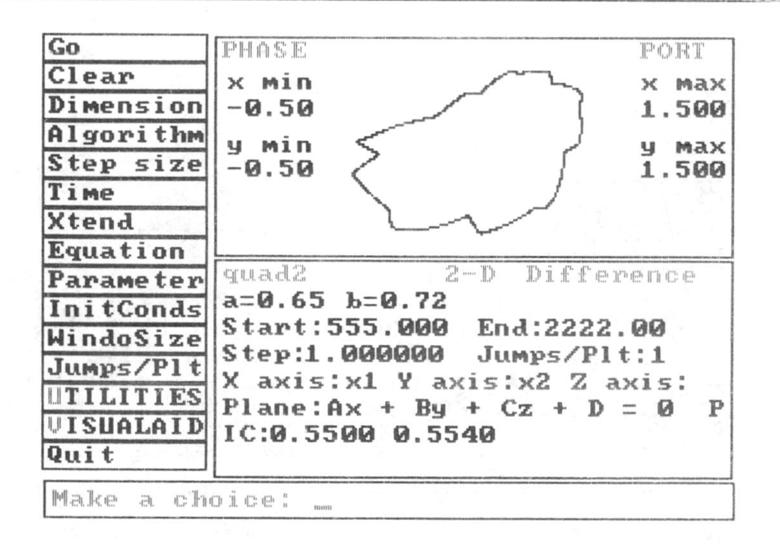

Figure 8.12. The invariant circle in *Figure 8.11* is about to break up. Through a complex sequence of bifurcations, the circle eventually gives rise to a strange attractor.

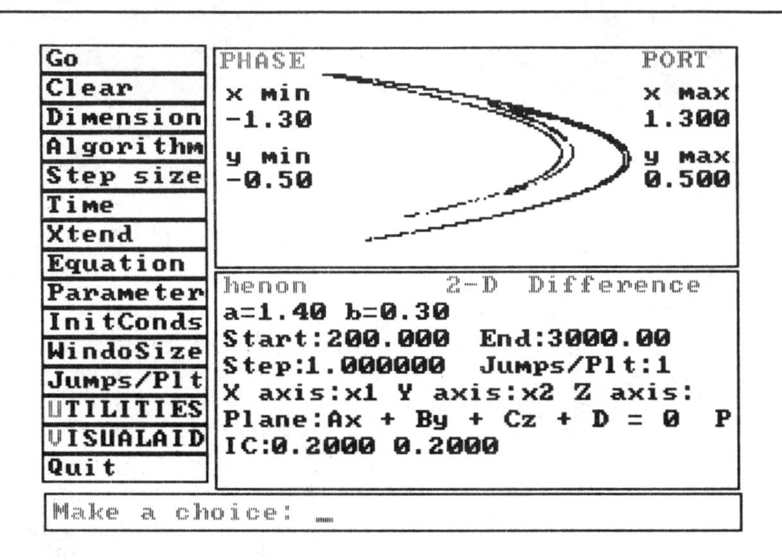

Figure 8.13. The "strange attractor" (proof awaiting) of the *henon* map.

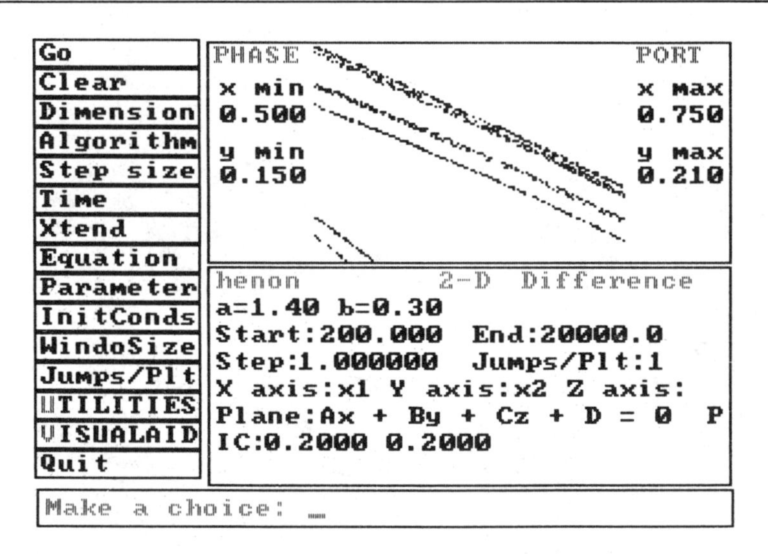

Figure 8.14. An enlarged piece of the Henon attractor in *Figure 8.13*, with the structure of a Cantor set cross an interval.

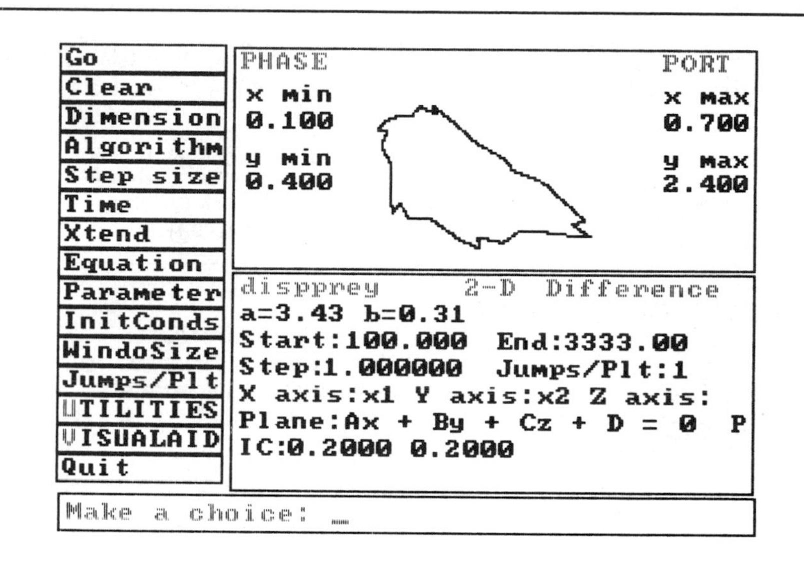

Figure 8.15. An invariant circle of the map *dispprey* is about to break up, eventually giving rise to the strange attractor below (compare with *Figure 8.12*).

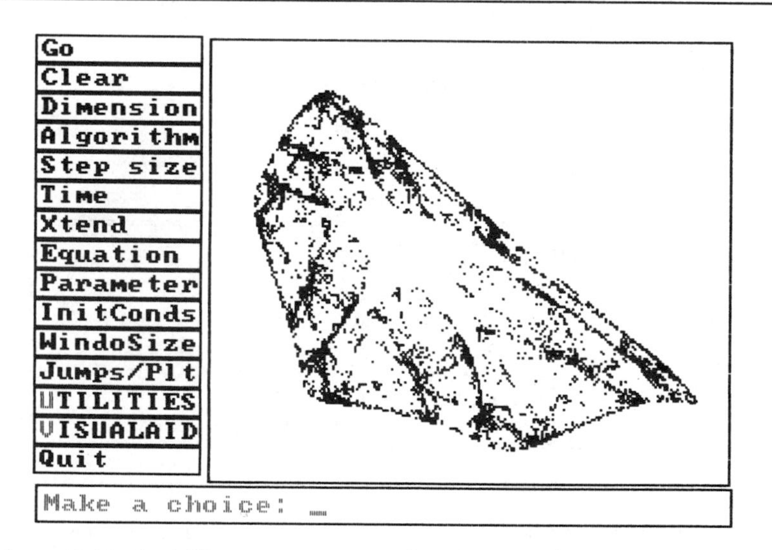

Figure 8.16. A visible strange attractor (proof awaiting) of the map *dispprey* at the parameter values $a = 3.65$ and $b = 0.31$.

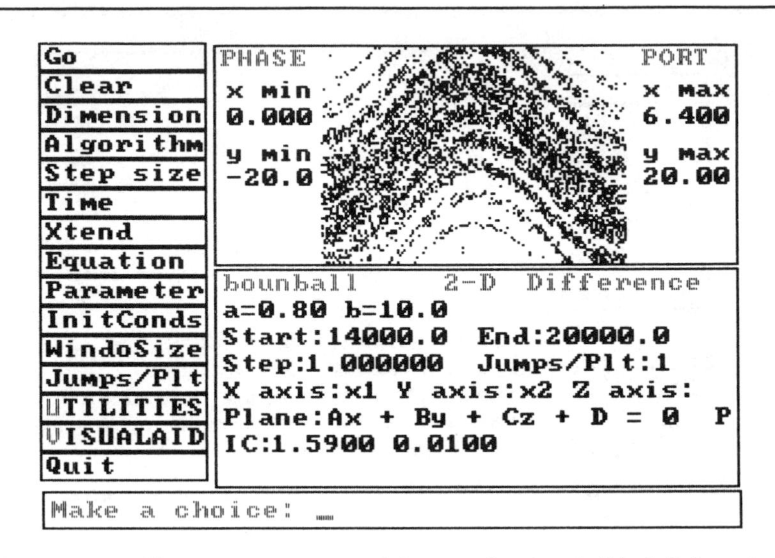

Figure 8.17. The strange attractor of the equation *bounball* (a ball bouncing on a periodically vibrating table).

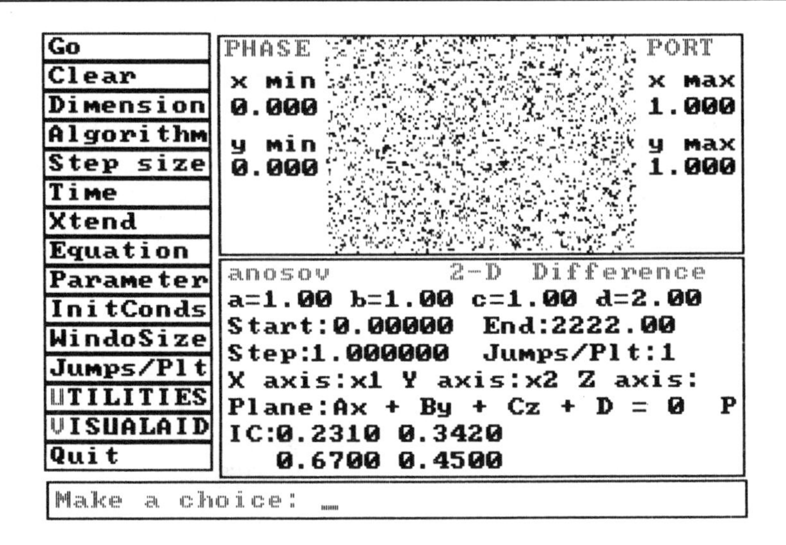

Figure 8.18. Two orbits of a toral automorphism, the equation *anosov*. Notice the ergodicity and mixing of the orbits.

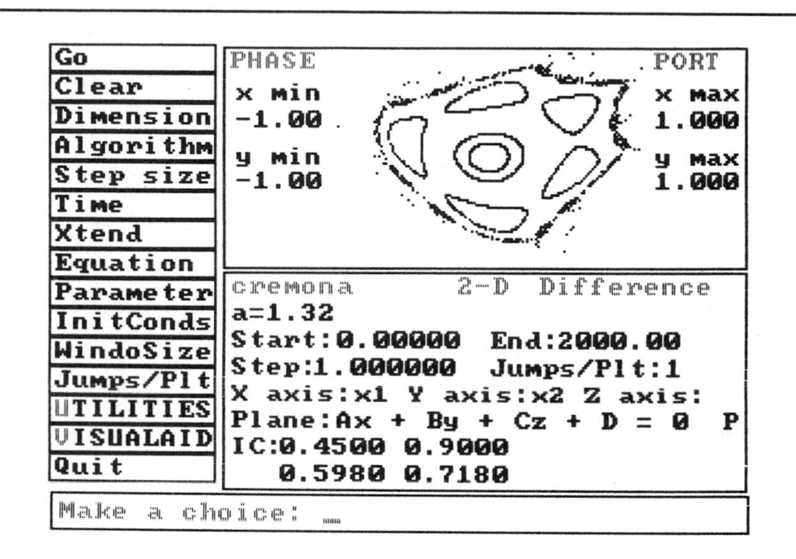

Figure 8.19. Three orbits of the *cremona* map near the elliptic fixed point at the origin.

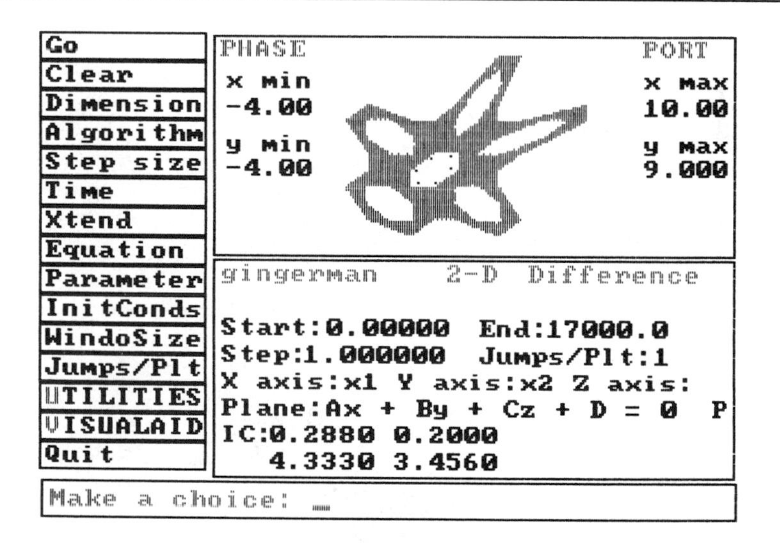

Figure 8.20. A periodic orbit of period six, and an ergodic orbit of the *gingerman* map near the elliptic fixed point at (1, 1).

8.3. 3D Difference Equations

When studying three-dimensional difference equations, you should use the *CutPlane* view with a vertical plane ($x_3 = constant$). This allows you to get a sense of depth, since parts of an orbit on different sides of a plane are plotted in different colors.

---

dislin3d *General three-dimensional discrete linear system.*

Type: 3D Difference.

Parameters: $a = 0.0$, $b = -1.0$, $c = 0.0$, $d = 1.0$, $e = 0.0$, $f = 0.0$, $g = 0.0$, $h = 0.0$, $i = -0.2$.

Equations:

$$x_1' = ax_1 + bx_2 + cx_3$$

$$x_2' = dx_1 + ex_2 + fx_3$$

$$x_3' = gx_1 + hx_2 + ix_3 .$$

The remarks for *dislin2d* apply to this system as well. You should investigate, by varying the parameters, all possible qualitatively different phase portraits of this system, as classified in the reference below.

Reference: Devaney [1985] p. 170.

---

act *A three-dimensional cubic map.*

Type: 3D Difference.

Parameters: $a = 0.6$, $b = 0.5$, $c = 0.3$, $d = 1.0$, $e = 1.0$.

Equations:

$$x_1' = ax_1 - b(x_2 - x_3)$$

$$x_2' = bx_1 + a(x_2 - x_3)$$

$$x_3' = cx_1 - dx_1^3 + ex_3 .$$

This three-dimensional map was introduced by **A**rneodo, **C**oullet, and **T**resser (hence the name) while they were studying the "screw-type" strange attractors of a one-parameter family of differential equations with a pair of Silnikov-type equilibria (see the equation *silnikov2*). As the parameter c is varied, this map exhibits many interesting phenomena including sequences of period-doubling bifurcations for stable invariant circles, strange attractors, etc.; see *Figures 8.21-24*.

References: Arneodo et al. [1981], [1982]; Du [1985].

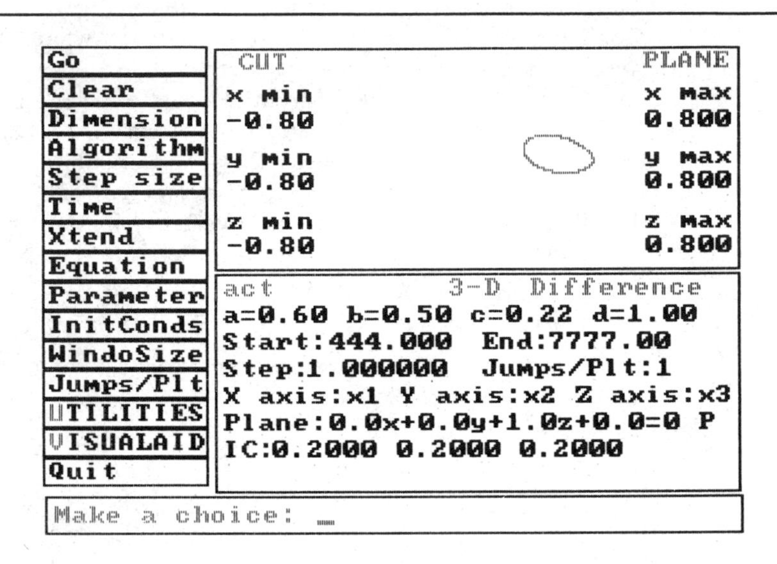

Figure 8.21. An attracting invariant circle of the three-dimensional map *act*. The next three figures show bifurcations of this circle.

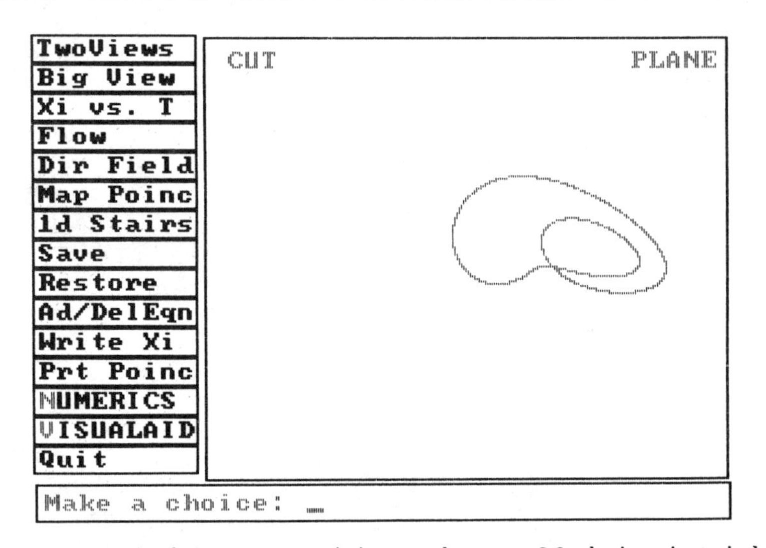

Figure 8.22. As the parameter c is increased to $c = 0.3$, the invariant circle in Figure 8.21 doubles.

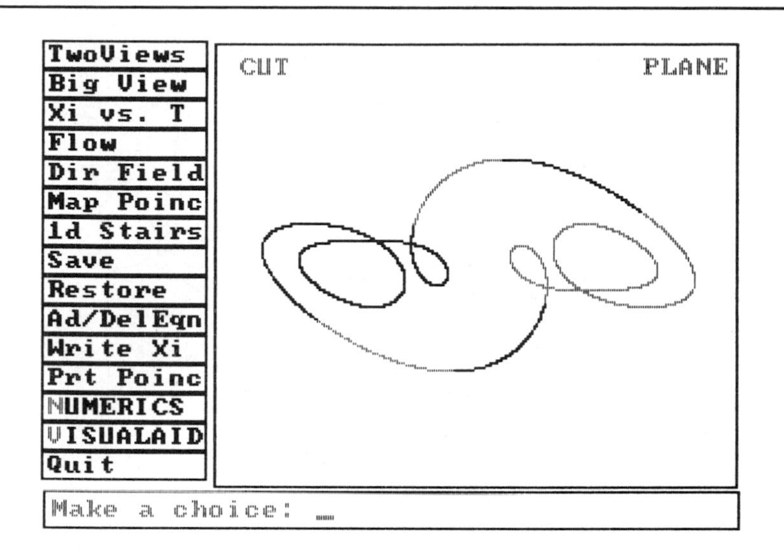

Figure 8.23. For the parameter value $c = 0.334$, the invariant circle of the map *act* doubles again (see Figures 8.21-22).

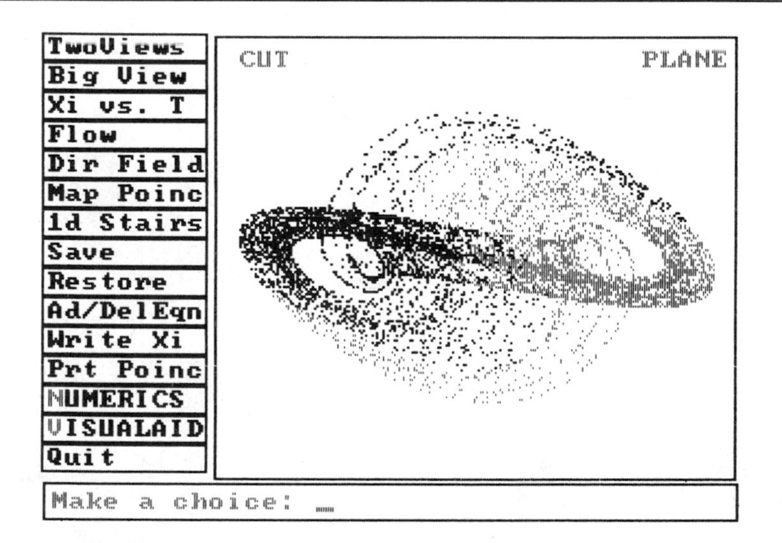

Figure 8.24. An apparent strange attractor of the map *act* for the parameter value $c = 0.376$ (compare with Figure 7.26).

What remains to be done ?

My main motivation for developing PHASER has been to lay a modest foundation for a "work station for experimental dynamics," one where students can not only get a taste of the current excitement in this field, but also perhaps discover new dynamical phenomena in an environment free from extensive programming efforts. Although the present version of PHASER has been successful in fulfilling my original goals, it has been difficult to resist adding new capabilities. However, as Hanns Sach has said, "an analysis terminates when the patient realizes that it could go on forever." Here is a list of additional utilities I personally would like to see:

- *More algorithms*: There is no "universal" numerical integration scheme that works well for all ordinary differential equations. Therefore, the most urgent addition is to provide a broader spectrum of algorithms, with features such as variable step size, variable order, implicit schemes, etc. Algorithms tailored to Hamiltonian systems, or to conservation of energy, momentum, etc. would also be desirable.

- *Equilibria and periodic orbits*: Just solving an initial value problem does not always provide the desired results in the elucidation of special features of phase portraits. Therefore, special methods for locating equilibrium points (a la Newton, for example) and for following periodic orbits are essential.

- *Invariant manifolds*: It is theoretically well known that many aspects of complicated dynamics can be understood in terms of the geometry of invariant manifolds. Unfortunately, locating such manifolds numerically in specific systems of ordinary differential equations is quite difficult, and new algorithms consequently need to be developed. This is a tractable problem in the case of difference equations, especially in two dimensions.

- *Power spectrum*: This is a useful tool in following bifurcations, especially in laboratory experiments. Routines incorporating fast Fourier transform algorithms should be provided so that the power spectra of orbits may be calculated (time series analysis).

- *Characteristic exponents, topological entropy, Hausdorff dimension, etc.*: These important numbers are useful measures of complexity in dynamical systems. Numerical schemes for calculating good approximations to these quantities should be developed.

- *4D graphics*: To produce mathematically informative images of an object in $\mathbf{R}^4$ on a two-dimensional screen is quite difficult. For example, the commonly used orthographic projection (see the *3dProject* entry on the utilities menu) introduces artificial singularities into the three-dimensional image of a torus in $\mathbf{R}^4$. To overcome some of these difficulties, four dimensional rotations, stereographic projection, etc. should be implemented.

To be realistic, the most commonly available configurations of personal computers are too restricted for the implementation of all of the features listed above. However, within the limitations of the hardware, the software, and the present design of PHASER, I hope to incorporate some of these ideas in future releases. In the meantime, I welcome further suggestions and criticisms of both PHASER and this book, as well as reports of any "bugs".

Appendix A

PHASER Quick Reference

To run your dynamical systems animator/simulator, type **phaser** and hit the **<return>** key. There are three main menus: *NUMERICS, VISUALAID*, and *UTILITIES*. Various submenus come up when further choices are appropriate. To activate a menu entry, press the first letter of the entry, but do not hit <return>. The only exception occurs when selecting from submenus of equations (which are sorted by *Dimension* and by differential/difference type determined by *Algorithm*), in which case you must type in the desired name in full, and press <return>. In addition, whenever user input is requested, you must type in the appropriate information, followed by <return>.

NUMERICS MENU	
Go	Compute and display. <esc> to halt temporarily.
Clear	Remove contents of graphical views.
Dimension	Specify dimension of equations to study.
Algorithm	Bring up a submenu of algorithms (*Difference* = maps).
Step Size	Set step size for Algorithm.
Time	Specify start and end of time for graphical output.
Xtend	Extend end time of computations.
Equation	Bring up a submenu of equations.
Parameter	Bring up a submenu of parameters in current equation.
InitConds	Specify (simultaneous) sets of initial conditions.
WindoSize	Adjust boundaries (scaling) of graphical views.
Jumps/Plt	Number of jumps (steps) between two plotted points.
UTILITIES	Bring up UTILITIES menu.
VISUALAID	Bring up VISUALAID menu.
Quit	Exit PHASER.

UTILITIES MENU	
Two Views	Choose two small views for the viewing area.
Big View	Choose one big view for the viewing area.
Xi vs T	Specify i of Xi and scaling for the view Xi vs. T.
Flow	Display flow on phase portrait view (in 2D only).
DirField	Display direction field (in 2D only).
MapPoinc	Set plane $Ax + By + Cz + D = 0$ and direction (N/P).
1d Stairs	Set power (iterate) of 1d map for stair step diagram.
Save	Save current screen in a file.
Restore	Restore a previously Saved screen.
Ad/delEqn	Add or delete equations to/from the library.
Write Xi	Write values of all (unrotated) Xi in a file.
PrtPoinc	Print values of (unrotated) Poincare map in a file.
NUMERICS	Bring up NUMERICS menu.
VISUALAID	Bring up VISUALAID menu.
Quit	Exit PHASER.

VISUALAID MENU	
Go	Compute and display. <esc> to halt temporarily.
Clear	Remove contents of graphical views.
3dProject	Specify a 3D projection from higher dimensions.
X-Rotate	Rotate current image about x-axis (in degrees).
Y-Rotate	Rotate current image about y-axis (in degrees).
Z-Rotate	Rotate current image about z-axis (in degrees).
EraseRota	Erase all rotations. Rotations are cumulative.
Body- Tog	Toggle body rotations on/off.
Axes- Tog	Toggle axes off/on.
RotAx-Tog	Toggle rotated axes on/off.
Flash-Tog	Toggle a flashing marker on/off.
Persp-Tog	Project from 3D to 2D using perspective.
NUMERICS	Bring up NUMERICS menu.
UTILITIES	Bring up UTILITIES menu.
Quit	Exit PHASER.

Using the *Two Views* and *Big View* entries on the UTILITIES menu, you can display any two small or one enlarged version of the nine views below. Their contents are:

VIEWS	
PhasePort	Phase portrait: for orbits, direction fields, and flows.
Set Up	Current settings of computational choices.
Xi vs T	Graph of variable Xi vs. time.
Equation	Text of current equations.
Last Xi	Last several (rotated) values of three projected variables.
CutPlane	Solutions in two colors on different sides of a plane.
MapPoinc	A planar Poincare map.
ValuPoinc	Last several (rotated) values of Poincare map.
1d Stairs	Stair step diagram for 1d difference equations (maps).

Appendix B

Library of PHASER

This appendix contains the list of all the difference and ordinary differential equations stored in the library of PHASER, along with their brief descriptions. Equations are sorted by *Dimension* and by difference/differential type, determined by *Algorithm*; see Lesson 8 of Chapter 5 on how to access a specific equation.

1D DIFFERENTIAL Equation	
cubic1d	One-dimensional cubic differential equation.

2D DIFFERENTIAL Equations	
linear2d	General two-dimensional linear system.
pendulum	Nonlinear pendulum on the plane.
predprey	Predator-prey, Volterra-Lotka, etc.
vanderpol	Oscillator of van der Pol -- A unique limit cycle.
saddlenod	Saddle-node -- Generic bifurcation of an equilibrium point.
transcrit	Transcritical bifurcation - Exchange of stability.
pitchfork	Bifurcation of an equilibrium with reflection symmetry.
hopf	Hopf bifurcation -- Birth of a periodic orbit.
dzero1	The generic unfolding of a double zero eigenvalue.
dzero2	Unfolding of a double zero eigenvalue with origin fixed.
dzero3	Unfolding of a double zero eigenvalue with symmetry.
hilbert2	A planar quadratic system with two limit cycles.
hilbert4	A planar quadratic system with four limit cycles.
averfvdp	Averaged forced van der Pol's oscillator.
MORE	To bring up the next page of equations.
gradient	The universal unfolding of the elliptic umbilic.

3D DIFFERENTIAL Equations	
lorenz	The most famous strange attractor.
linear3d	General three-dimensional linear system.
vibration	Periodically forced linear oscillator.
bessel	Bessel's equation.
euler	Euler's equation.
laguerre	Laguerre's equation.
legendre	Legendre's equation.
forcevdp	Periodically forced van der Pol's oscillator.
forcepen	Periodically forced pendulum.
mathieu	Mathieu's equation.
forceduf	Periodically forced Duffing's equation.
rossler	A not-so-strange strange attractor.
zeroim	Unfolding of zero and pure imaginary eigenvalues.
silnikov	A homoclinic orbit and horseshoes in three dimensions.
MORE	To bring up the next page of equations.
silnikov2	A pair of Silnikov-like homoclinic orbits.

4D DIFFERENTIAL Equations	
harmoscil	A pair of linear harmonic oscillators.
kepler	Kepler and anisotropic Kepler problems.
r3body	The restricted problem of three bodies on the plane.
henheile	A non-integrable Hamiltonian by Henon & Heiles.
coplvdp1	Two coupled van der Pols -- An attracting torus.
coplvdp2	Two coupled van der Pols -- Two invariant tori.
couplosc	Coupled oscillators.
reson21	Normal form of 2:1 resonance (non-hamiltonian).

1D DIFFERENCE Equations	
logistic	The logistic map -- The one that started it all.
dislin1d	General one-dimensional linear map.
discubic	General one-dimensional cubic map.
newton	Calculating square roots with Newton-Raphson.
tent	Piecewise linear version of the logistic map.
singer	A one-hump map with two attractors.
mod	Linear modulus map -- A pseudo-random number generator.
arnold	The standard circle map.

2D DIFFERENCE Equations	
dislin2d	General two-dimensional discrete linear system.
gauss	Fast computation of elliptic integrals.
quad1	A Henon-like quadratic map on the plane.
quad2	A quadratic map exhibiting Hopf bifurcation.
henon	A quadratic map on the plane with a strange attractor.
lozi	Piecewise linear version of the Henon map.
dellogis	Two-parameter delayed logistic map.
dispprey	Discrete predator prey model.
bounbal	Bouncing ball on a periodically vibrating table.
anosov	Ergodic toral automorphism of Anosov.
cremona	An area-preserving quadratic (standard) map.
icremona	The inverse of cremona map.
gingerman	A piecewise linear area-preserving map.

3D DIFFERENCE Equations	
dislin3d	General three-dimensional discrete linear system.
act	A three-dimensional cubic map.

References

This is a selective annotated list of references directly related to the equations stored in the library of PHASER. They have been chosen to provide a point of entry into the vast literature on differential and difference equations. They are therefore not necessarily the original sources where a particular equation first appeared. I apologize for the inevitable omissions, and urge you to consult some of the entries below for the general theory as well as further references. Boyce & DiPrima [1977] and Braun [1983] are standard for beginners. Arnold [1973], Arrowsmith & Place [1982], Devaney [1985], and Hirsch & Smale [1973] are good intermediate-level texts. Finally, recent books by Chow & Hale [1982] and Guckenheimer & Holmes [1983], and the review article by Whitley [1983] are the advanced sources of choice.

Books

ABRAHAM, R. and SHAW, C. [1982]. *Dynamics, the geometry of behavior. Part one: Periodic behavior; Part two: Chaotic behavior; Part three: Global behavior.* Aerial Press: P.O. Box 1360, Santa Cruz, CA 95061.

A profusely illustrated popular account.

ARNOLD, V.I. [1983]. *Geometric methods in the theory of ordinary differential equations.* Springer-Verlag: New York, Berlin, Heidelberg.

The inside story from a master; devoid of formalities.

ARNOLD, V.I. [1978]. *Mathematical methods of classical mechanics.* Springer-Verlag: New York, Berlin, Heidelberg.

A unique blend of old and new views on this time-honored subject.

ARNOLD, V.I. [1973]. *Ordinary differential equations.* M.I.T. Press: Cambridge, London.

By a master; on the shape of things to come in beginning courses in ordinary differential equations.

ARNOLD, V.I. and AVEZ, A. [1968]. *Ergodic problems of classical mechanics.* Benjamin: New York, Amsterdam.

Another masterpiece by Arnold, with a marvelous collection of appendices.

ARROWSMITH, D.K. and PLACE, C.M. [1982]. *Ordinary differential equations: A qualitative approach with applications.* Chapman and Hall: London, New York.

Coverage similar to that of Hirsch & Smale, but no proofs.

BOYCE, W.E. and DiPRIMA, R.C. [1977]. *Elementary differential equations and boundary value problems, Third Edition.* John Wiley & Sons: New York.

The most popular elementary introduction to the subject. Many good problems and applications.

BRAUN, M. [1984]. *Differential equations and their applications, Third Edition.* Springer-Verlag: New York, Berlin, Heidelberg.

Another popular introductory text, with a nice treatment of qualitative theory. Has realistic case studies in modeling using ordinary differential equations.

CARR, J. [1980]. *Applications of center manifold theory.* Springer-Verlag: New York, Berlin, Heidelberg.

A readable introduction to one of the key ideas in bifurcation theory. Nice examples.

CHOW, S.N. and HALE, J.K. [1982]. *Methods of bifurcation theory.* Springer-Verlag: New York, Berlin, Heidelberg.

A state-of-the-art treatise on various aspects of bifurcation theory.

COLLET, P. and ECKMANN, J.-P. [1980]. *Iterated maps on the interval as dynamical systems.* Birkhauser: Boston.

More than you may want to know about interval maps.

CONTE, S.D. and DeBOOR, C. [1972]. *Elementary numerical analysis, Second Edition.* McGraw-Hill: New York.

A rigorous introductory text, with Fortran programs.

DEVANEY, R.L. [1985]. *An introduction to chaotic dynamical systems.* Benjamin-Cummings: Menlo Park, CA.

An elementary introduction to modern dynamics through maps.

FOLEY, J.D. and VAN DAM, A. [1982]. *Fundamentals of interactive computer graphics.* Addison-Wesley: Reading, MA.

The book on computer graphics.

FORSYTHE, G.E., MALCOLM, M.A., and MOLER, C.B. [1977]. *Computer methods for mathematical computations.* Prentice-Hall: Englewood Cliffs, N.J.

A very readable survey on numerical methods for use in scientific computations.

GEAR, C.W. [1971]. *Numerical initial value problems in ordinary differential equations.* Prentice-Hall: Englewood Cliffs, New Jersey.

An advanced classic on the subject.

GRENANDER, U. [1982]. *Mathematical experiments on the computer.* Academic Press: New York, London.

An unusual book, arguing in favor of using "the computer as the mathematician's laboratory". We obviously agree.

GUCKENHEIMER, J. and HOLMES, P.J. [1983]. *Nonlinear oscillations, dynamical systems and bifurcations of vector fields.* Springer-Verlag: New York, Berlin, Heidelberg.

A good survey of local and global aspects of modern dynamical systems. Nice examples.

HALE, J.K. [1963]. *Oscillations in nonlinear systems.* McGraw-Hill: New York, Toronto, London.

An influential book in this revitalized area.

HALE, J.K. [1969]. *Ordinary differential equations.* Wiley-Interscience: New York, London.

A standard on advanced theory of ordinary differential equations.

HASSARD, B.D., KAZARINOFF, N.D., and WAN, Y.-H. [1981]. *Theory and application of Hopf bifurcation.* Lon. Math. Soc. Lecture Notes **41.**

A computer-aided determination of stability in Hopf bifurcation.

HIRSCH, M. and SMALE, S. [1974]. *Differential equations, dynamical systems and linear algebra.* Academic Press: New York, San Francisco, London.

A nice book after Boyce & DiPrima or Braun, provided you do not get stuck on the linear algebra, and ignore the misprints. A broad spectrum of applications.

KNUTH, D.E. [1980]. *The art of computer programming, Volume 2: Seminumerical algorithms.* Addison-Wesley: Reading, MA.

A seven-volume series defining theoretical computer science.

LICHTENBERG, A.J. and LIEBERMAN, M.A. [1983]. *Regular and stochastic motion.* Springer-Verlag: New York, Berlin, Heidelberg.

Mathematical style is somewhat hasty, but a useful source on "conservative chaos".

MARSDEN, J.E. and McCRACKEN, M. [1976]. *The Hopf bifurcation and its applications.* Springer-Verlag: New York, Berlin, Heidelberg.

Another book devoted to this most celebrated bifurcation and its ramifications, including an English translation of Hopf's original paper.

MAYNARD SMITH, J. [1968]. *Mathematical ideas in biology.* Cambridge University Press: London, New York.

A nice introduction to theoretical biology, one which has been a rich source of discrete systems.

SIEGEL, C.L. and MOSER, J.K. [1971]. *Lectures on celestial mechanics.* Springer-Verlag: New York, Berlin, Heidelberg.

Two master mathematicians' account of celestial mechanics.

SIMMONS, G.F. [1972]. *Differential equations with applications and historical notes.* McGraw-Hill: New York, San Francisco.

A nonstandard book on a very standard subject. Delightful style with historical notes.

SPARROW, C. [1982]. *The Lorenz equations: Bifurcations, chaos and strange attractors.* Springer-Verlag: New York, Berlin, Heidelberg.

A good case study on how to dissect a specific set of equations with the help of both theory and numerical computations. A comprehensive guide to the most famous equations in our field in recent years.

YEH, Y.C. [1986]. *Theory of limit cycles.* American Math. Soc.: Providence, Rhode Island.

How difficult can a quadratic planar system be?

Articles

AIZAWA, Y. and SAITO, N. [1972]. "On the stability of isolating integrals," *J. Phys. Soc. Jap.*, **32**, 1636-1640.

ARNEODO, A., COULLET, P., and TRESSER, C. [1981]. "Possible new strange attractors with spiral structure," *Commun. Math. Phys.*, **79**, 573-579.

[1982]. "Oscillators with chaotic behavior: An illustration of a theorem of Shil'nikov," *J. Stat. Phys.*, **27**, 171-182.

ARNOLD, V. [1965]. "Small denominators, I: Mappings of the circumference onto itself," *AMS Trans. Ser. 2*, **46**, 213-284.

ARONSON, D.G., CHORY, M.A., HALL, G.R., and McGEHEE, R.P. [1982]. "Bifurcations from an invariant circle for two-parameter families of maps of the plane: A computer assisted study," *Commun. Math. Phys.*, **83**, 303-354.

ARONSON, D.G., DOEDEL, E.J., and OTHMER, H.G. [1985]. "An analytic and numerical study of the bifurcations in a system of linearly coupled oscillators," submitted to *Physica D.*

BAMON, R. [1984]. "Solution of Dulac's problem for quadratic vector fields," Preprint from I.M.P.A.

BAXTER, R., EISERIKE, H., and STOKES, A. [1972]. "A pictorial study of an invariant torus in phase space of four dimensions," in *Ordinary Differential Equations*, edited by L. Weiss, 331-349.

BORWEIN, J.M. and BORWEIN, P.B. [1984]. "The arithmetic-geometric mean and fast computation of elementary functions," *SIAM Review*, **26**, 351-366.

BOUNTIS, T.C., SEGUR, H., and VIVALDI, F. [1982]. "Integrable Hamiltonian systems and the Painleve property," *Phys. Rev.*, **A25**, 1257.

CASASAYAS, J. and LLIBRE, J. [1984]. "Qualitative analysis of the anisotropic Kepler problem," *Memoirs Am. Math. Soc.*, **312**.

CHENCINER, A. [1983]. "Bifurcations de diffeomorphismes de $\mathbf{R}^2$ au voisinage d'un point fixe elliptique," in *Les Houches, Session XXXVI, 1981 - Chaotic behavior of deterministic systems*, edited by G. Iooss, R.H.G. Helleman and R. Stora, North-Holland Publishing Company.

CHICONE, C. and TIAN, J.H. [1982]. "On general properties of quadratic systems," *Amer. Math. Monthly*, **89**, 167-179.

CHURCHILL, R., PECELLI, G., and ROD, D. [1979]. "A survey of the Henon-Heiles Hamiltonian with applications to related examples," *Springer Lecture Notes in Physics*, **93**, 76-136.

CURRY, J. [1979]. "On the Henon transformation," *Commun. Math. Phys.*, **68**, 129-140.

DEVANEY, R. [1982]. "Blowing up singularities in classical mechanics," *Amer. Math. Monthly*, **89**, 535-551.

[1984]. "A piecewise linear model for the zones of instability of an area preserving map," *Physica 10D*, 387-393.

DEVANEY, R. and NITECKI, Z. [1979]. "Shift automorphisms in the Henon mapping," *Commun. Math. Phys.*, **67**, 137-148.

DU, BAU-SEN [1985]. "Bifurcation of periodic points of some diffeomorphisms on R^3," *Nonlinear Analysis*, **9**, 309-319.

GUCKENHEIMER, J. [1984]. "Multiple bifurcation problems of codimension two," *SIAM J. Math. Anal.*, **15**, 1-49.

GUTZWILLER, M. [1973]. "The anisotropic Kepler problem in two dimensions," *J. Math. Phys.*, **18**, 139-152.

HENON, M. [1969]. "Numerical study of quadratic area-preserving mappings," *Quart. Appl. Math.*, **27**, 291-312.

[1976]. "A two dimensional mapping with a strange attractor," *Comm. Math. Phys.*, **50**, 69-77.

[1983]. "Numerical exploration of Hamiltonian systems," in *Les Houches, Session XXXVI, 1981 - Chaotic behavior of deterministic systems*, edited by G. Iooss, R.H.G. Helleman and R. Stora, North-Holland Publishing Company.

HENON, M. and HEILES, C. [1964]. "The applicability of the third integral of motion: Some numerical experiments," *Astron. J.*, **69**, 73-79.

HERMAN, M. [1979]. "Sur la conjugation differentiable des diffeomorphismes du cercle a des rotations," *Publ. Math. IHES*, **49**.

HOLMES, P.J. and RAND, D. [1978]. "Bifurcations of the forced van der Pol oscillator," *Quart. Appl. Math.*, **35**, 495-509.

KOCAK, H., BISSHOPP, F., BANCHOFF, T., and LAIDLAW, D. [1983]. "Linear oscillators and the hypersphere," Computer generated color film, Brown University.

[1986]. "Topology and mechanics with computer graphics: Linear Hamiltonian systems in four dimensions," *Advances in Applied Mathematics*, 7, 282-308.

KOCAK, H., MERZBACHER, M., and STRICKMAN, M. [1984]. "Dynamical systems with computer experiments at the Brown University Instructional Computing laboratory," preprint, Brown University.

LANFORD, O. [1980]. "Smooth transformations of intervals," *Seminaire Bourbaki*, **33e annee**, no.563.

[1982]. "A computer assisted proof of the Feigenbaum conjecture," *Bull. Amer. Math. Soc.*, **6**, 427-434.

LANGFORD, W.F. [1982]. "Chaotic dynamics in the unfolding of degenerate bifurcations," Report No. 82-2, Department of Mathematics, McGill University.

[1985]. "Unfolding of degenerate bifurcations," in *Chaos, Fractals and Dynamics*, edited by P. Fisher and W.R. Smith, Marcel-Dekker.

LI, T-Y. and YORK, J. [1975]: "Period three implies chaos," *Amer. Math. Monthly*, **82**, 985-992.

LORENZ, E. [1963]. "Deterministic non-periodic flows," *J. Atmos. Sci.*, **20**, 130-141.

LOZI, R. [1978]. "Un attracteur etrange(?) du type attracteur de Henon," *J. Phys.*, **39**, 9-10.

MAY, R. [1976]. "Simple mathematical models with very complicated dynamics," *Nature* **261**, 459-467.

MILNOR, J. [1983]. "On the geometry of the Kepler problem," *Amer. Math. Monthly*, **90**, 353-364.

MISUREWICZ, M. [1980]. "Strange attractors for the Lozi mapping," *Nonlinear dynamics*, Annals of the N. Y. Academy of Sciences, **357**, 348-358.

MOSER, J.K. [1968]. "Lectures on Hamiltonian systems," *Memoirs Am. Math. Soc.*, **81**.

NEWHOUSE, S.E. [1980]. "Lectures on dynamics," *Progress in Mathematics*, **8**, Birkhauser: Boston, 1-114.

NITECKI, Z. [1981]. "Topological dynamics on the interval," in *Ergodic theory and dynamical systems*, Birkhauser: Boston.

POUNDER, J.R., ROGERS, T.D. [1980]. "The geometry of chaos: dynamics of a nonlinear second order difference equation," *Bull. Math. Biol.*, **42**, 551-597.

RAND, D.A. [1983]. "Universal properties and renormalization in dynamical systems," Notes from International Centre for Theoretical Physics, Trieste, Italy.

ROGERS, T.D. and WHITLEY, D.C. [1983]. "Chaos in the cubic mapping," *Math. Modelling*, **4**, 9-25.

SAARI, D.G. and URENKO, J.B. [1984]. "Newton's method, circle maps, and chaotic motion," *Amer. Math. Monthly*, **91**, 3-18.

SHI, S.-L. [1980]. "A concrete example of the existence of four limit cycles for plane quadratic systems," *Sci. Sinica*, **23**, 153-158.

SILNIKOV, L.P. [1965]. " A case of the existence of a denumerable set of periodic motions," *Sov. Math. Dokl.*, **6**, 163-166.

SINGER, D. [1978]. "Stable orbits and bifurcations of maps of the interval," *SIAM J. Appl. Math.*, **35**, 260-268.

TAKENS, F. [1974]. "Forced oscillations and bifurcations," *Comm. Math. Inst. Rijkuniversiteit Utrecht*, **3**, 1-59.

ULAM, S.M., von NEUMANN, J. [1947]. "On combinations of stochastic and deterministic processes," *Bull. Amer. Math. Soc.*, **53**, 1120.

WHITLEY, D. [1983]. "Discrete dynamical systems in dimensions one and two," *Bull. London Math. Soc.*, **15**, 177-217.

YEH, Y.C. [1958]. *J. Nanjing Univ.*, **1**, 7-17.

[1982]. "Problems in the qualitative theory of ordinary differential equations," *J. Differential Equations*, **46**, 153-165.

Index

Menu entries and equation names are in **bold face.**